THE MARTIAN ALMANAC

FOR THE MARTIAN YEAR 221

VOLUME 1

2024 NOVEMBER 11 - 2025 OCTOBER 20

THOMAS GANGALE

Printing history

1st edition December 2022
2nd edition April 2024

TABLE OF CONTENTS

To all who have ever believed in Mars, whether as a source of protection, as an abode of life, as a haven for the soul, as the target of exploration, or as the destiny of humankind.

We are the Martians.

The Martians stared back up at them for a long, long silent time from the rippling water....

—Ray Bradbury, "The Million Year Picnic," 1946

THE MARTIAN TIME-SLIP CONTINUES

Along with Martian the release of literature, art, music, film and of plans for human Mars mission of the late Darian second century, together with the birthdates of people associated with these events, the Martian Almanac for the marsrev 221 documents the first full calendar marsrev of Mars Exploration Rover operations on the surface of Mars. The journeys taken by *Spirit* and *Opportunity* initiated a period of continuous exploration of the Martian surface; consequently, the content of the 221 Almanac is more heavily weighted toward events on Mars in contrast to the 220 Almanac, which largely documented Martian events on Earth. As of this writing, we are still in this period of continuous surface exploration, thus succeeding editions of the Almanac will likewise report the sol-by-sol activities of vehicles on the surface.

On a technical note regarding late second century telescopic observations of Mars, in comparing drawings based on these observations with computer-generated representations of the aspect that Mars presented to Earth, it appears that various sources such as the *Annals of the Lowell Observatory* and Pickering's reports on Mars in *Popular Astronomy* used different schemata in recording times of observations. At times, observations were recorded with reference to midnight Greenwich Mean Time (time zone Z), while other observa5tions appear to have referenced noon Greenwich Mean Time (Z+12, or time zone Z advanced 12 hours, a common practice among astronomers of the time, with the date beginning at noon so that an entire night's observations would bear the same date), while still others are best understood as being referenced to noon Arizona time (T+12, or time zone T advanced 12 hours).

This second edition corrects mistakes which appeared in the first edition, in which Michael Allison's Mars Sol Date (MSD) and the Curiosity Sol were off by one count. My thanks to Dr. Allison for pointing out these errors. Also, the Zhurong Sol counts have been deleted, as this mission ended before the end of marsrev 219. The computer displays of Mars, which serve to better understand the drawings made by early astronomers, have been upgraded from monochrome to color, and the terminator is displayed (this was redacted in the earlier monochrome displays to mask a software error).[1] Finally, the imagery from the 1974 *Mars 4* mission have been correlated to the 2021 United States Geological Survey's Mars Chart series to given then context.

[1] These computer displays have a retro feel to them for good reason. I originally developed the software in the late 1980s in a Microsoft QuickBASIC environment on a PC clone with an 8-bit 8088 processor, 8087 math coprocessor, 640K RAM, 4.77 MHz clock speed, Enhanced Graphics Array with 640 x 480 screen resolution, and dual 720K disk drives, running MS-DOS 2.1. Getting into the 35-year-old source code required getting a vintage Windows 98 laptop. This also enabled me to save screen captures of the EGA displays; the monographic displays in the first edition were digital camera images of the screen on a later-model computer running the executable code… a kludge solution.

MARTIAN CALENDAR TIME

Time has a texture in human perception; it has a fragrance and a taste, and in it are embedded shapes, colors, sounds, and emotions. A year is not merely a number; it brings to mind where we were, what we were doing, who was with us, and what music was popular during that time. The difference between one year and another is not a simple mathematical operation of subtraction, it is a complex navigation through human experiences. Even if we were not alive during a certain year—1945, 1776, 1492—by virtue of a sense of history it has a meaning to us that evokes a flood of sensations, and on specific dates as we voyage through each year—July 4, September 11, December 7—we may observe anniversaries with pride, or in sorrow, or as one which lives in infamy.

1969 July 20 is a date that belongs more to the Moon than to Earth. Mars too has its own set of dates, although none so celebrated as this one, since a human boot print has yet to be impressed upon its ruddy regolith, but they exist nonetheless—notes and sketches taken down at observatories, stories and illustrations published in magazines and books, music released on vinyl and optical disc—and as that world becomes a new stage peopled with players yet unborn, new layers of texture will be laid upon Martian time, becoming ever richer and more vibrant. As *Ecclesiastes* 3:1 reminds us, "To everything there is a season, and a time for every purpose;" just as a particular month on Earth brings to mind summer breezes or winter storms and how they bear on human activities, so future poets will write of Martian months and seasons, expanding on the brief glimpse of Oliver Wendell Holmes from millions of miles away:

> The snows that glittered on the disk of Mars
> Have melted, and the planet's fiery orb
> Rolls in the crimson summer of its year;

As to the practical uses of marking the passage of time, we live the present day by day, waking, working, eating, recreating, sleeping, et cetera; furthermore, we reflect on the past and plan for the future in terms of weeks, months, and years. The day and the year are astronomical cycles, based respectively on the rotation of Earth on its axis and its orbit around the Sun. The month is also an astronomical cycle in lunar calendars (for instance, the Islamic calendar) and lunisolar calendars (such as the Chinese and Hebrew calendars), being based on the synodic period of the Moon around Earth of 29.53 days, but in solar calendars (for example, the Gregorian calendar), the month is a socially-constructed unit of time dividing the year into twelve parts; likewise, the week, although originally suggested by the four principal phases of the Moon (new moon, first quarter, full moon, and third quarter), is also a social construct of seven days.

We will transplant this legacy set of practices to Mars and adapt it to the planet's distinct natural cycles. A lucky break for humans is that the rotation of Mars is not greatly different from that of Earth, thus our sleep/wake cycle of 24 hours will not be thrown off kilter by a 10-hour or 40-hour rotation period as we may struggle to contend with on habitable extrasolar planets in the far future. On Mars, the solar day/night cycle, known as the "sol," is 24.67 hours, well within proven human adaptability. Because the rotational axis of Mars is tilted at a similar angle to the plane of its orbit (25.2° as compared to Earth's 23.4°), the Martian environment experiences pronounced seasonal changes, which will have important economic ramifications such as power generation (both solar and wind) and consumption (especially heating), thus the period of Mars's revolution around the Sun must be taken into account in a timekeeping system for human activities on that planet.[2] Having nothing at all to do with Mars's astronomical cycles, the month and the week are so deeply rooted in the human non-material culture that they ought to be adapted to Mars, absent a compelling reason not to do so. It is upon these basic principles that I developed the Darian calendar for Mars in 1985, published in the *Journal of the British Interplanetary Society* the following year. It is certainly not the first Martian calendar ever invented, and many others have been proposed since, but it is one that has come to be widely diffused and used, having been mentioned in at least four novels set on Mars, printed and sold as calendars in several countries and languages, and implemented in about a half-dozen web applications; there are even plans to market Darian timepieces. Meanwhile, since 2004, JPL teams have been living according to the time on Mars as they have continuously operated the various rovers and stations on the Martian surface. As the exploration of Mars by remote presence continues, leading to eventual human expeditions, stations, and settlements, the desirability of a diary of events pertaining to the history of Mars, combined with the need emerges for a "sol planner" for rover and station operators as well as for the interested public.

The design of the Darian calendar is as follows. There are 668.6 sols in a Martian revolution around the Sun; since no distinctive term is now in common use for this period of time, as "sol" is distinct from the Earth "day," this volume uses the term "marsrev". Just as there are longer and shorter calendar years in Earth's Gregorian calendar to keep it in synchronization with the 365.25-day year (the longer year is referred to informally as a leap year, formally as a bissextile year[3]), there are longer and shorter calendar marsrevs; however, in reverse of the Gregorian calendar, on Mars, the longer year is more common than the shorter year, in other words, a leap backward rather than forward. Because of this, the occasional shorter year is referred to as "demosextile" since a diurnal period is deducted *(demo)* rather than repeated *(bis)*. A common calendar marsrev contains 669 sols; a demosextile marsrev contains 668 sols. Demosextile marsrevs are those whose sequential number is divisible by 2, except for those which are divisible by 10; thus, there are six common years in every decade. For example, the marsrev 218 is demosextile because it is divisible by 2, 219 is common because it is not divisible by 2, and 220 is common because it is divisible by 10. The calendar marsrev begins on or about the sol when Mars is at solar longitude (L_S) = 0°, which is the vernal equinox in the northern hemisphere and the autumnal equinox in the southern hemisphere; to eliminate the need to specify which hemisphere is at issue, the terms northward equinox is used since the Sun is crossing Mars's celestial equator from south to north. Similarly, the other seasons of Mars are marked by the northern solstice, the southward equinox, and the southern solstice.[4]

---

[2] This already occurs; the National Aeronautics and Space Administration's reports of *Spirit* and *Opportunity* rover activity document restrictions in activity around the winter solstice due to low Sun angles reducing available power.

[3]The term derives from the Romans practice of repeating *ante dies VI Kalendas Martius,* "the sixth day before the Kalends of March."

[4] Not only does Mars's axis have a slightly different inclination, it also is tilted in a much different direction. The northern solstice occurs when the Sun is in the constellation Pisces as seen from Mars, while the constellation of the southern solstice is Virgo; thus, the northern tropic circle is Tropicus Piscium and the southern tropic circle is Tropicus Virginis.

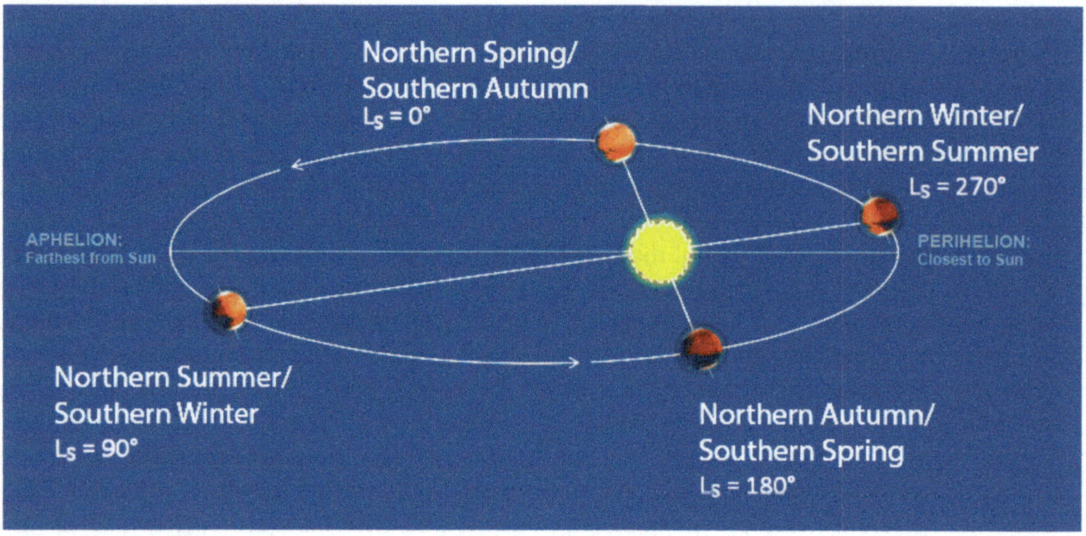

The length of the calendar month is close to that of the month on Earth; again, no distinct term for this unit of time is in common use, but given that language for Martian surface feature nomenclature is Latin, the Martian analog of the month might be called the "mensis" similarly, the Martian analog of the week might be called the "vicis".[5] In any case, there are 24 menses, typically comprised of 28 sols, into which the 7-sol vicis fits perfectly. Well, almost perfectly. In demosextile marsrevs, four of the menses, spaced evenly throughout the marsrev, contain only 27 sols. Each mensis begins on the first day of the vicis. The menses are named for twelve of the constellations through which the Sun passes along the Martian ecliptic during the course of the marsrev,[6] twelve Latin names alternating with twelve Sanskrit names in a blending of western and eastern cultural influences.[7] The sols of the vicis are taken from the Latin names of the terrestrial days of the week, most of which will be familiar to the 800 million speakers of Romance languages.[8]

At this point, other Martian timekeeping conventions deserve mention. NASA, including JPL, counts sols from the beginning of each mission on the surface of Mars; this practice began with the landing of *Viking 1* on 1976 July 20, although the practice has been

---

[5] The terms "week" and "month" tend to be used with less precise intent than are "day" and "year," rather like "further" is used more metaphorically whereas "farther" is meant more literally. For this reason, as well as the fact that the length of a Martian week or month is not radically different from its Earthly counterpart, distinctive Martian terms may never come into common usage.

[6] The first known mention of Ophiuchus, the thirteenth constellation, sandwiched between Scorpius and Sagittarius, was not until the 4th century BCE, by which time the tradition of twelve zodiacal constellations, descending from Babylonian astronomy, had been established for a millennium.

[7] The Sanskrit names for the twelve ancient zodiacal constellations, or derivations thereof, became the names of the months in the Hindu solar calendar, representing the time of year during which the Sun passed through these constellations. The correlation between the Sun's passage and the timing of the months was inexact from the beginning, as the months were approximately equal spans of time (30 or 31 days), therefore representing approximately equal arcs of 30 degrees along the ecliptic; however, the zodiacal constellations do not span equal arcs, with some being wider and some narrower. As the millennia have passed, the correlation between the Sun's passage and the timing of the months on the Hindu solar calendar has been entirely broken due to the precession of Earth's rotational axis.

[8] Sol Solis, Sol Lunae, Sol Martis, Sol Mercurii, Sol Jovis, Sol Veneris, Sol Saturni.

inconsistent in that some missions have begun with Sol 0 while others have begun with Sol 1. In 1998, planetary scientist Michael D. Allison proposed the Mars Sol Date (MSD) as a Martian analog to the terrestrial Julian Day; the epoch he chose was a date on which there was "a near coincidence of the Earth and Mars planetocentric solar longitudes," 1873 December 29. In a 2000 peer-reviewed journal article, planetary scientists Robert T. Clancy *et alia* adopted the arbitrary convention in which 1955, April 11, when Mars was at L_S=0°, was the beginning of Martian year 1. These conventions make perfect sense from the focus of mission controllers and planetary scientists in their work; the Darian calendar, however, is intended as a civil calendar for an emergent Martian culture, and the civil calendar counts years in celebration of some event of great cultural importance. For instance, a common practice in ancient times was to count the years from the beginning of the reign of the current monarch; the epoch for the Hebrew calendar is the date of Creation (3761 BCE) as calculated by the medieval Sephardic Jewish philosopher Moses ben Maimon (Maimonides); various epochs for the Chinese calendar are defined by the beginning of the reign of an emperor (Huángdì, 2697 BCE; Yáo, 2156 BCE) or the birth year of Kǒng Fūzǐ (Confucius, 551 BCE); the Romans counted years from the founding of their city (753 BCE), the Gregorian calendar from the birth of Jesus, and the Islamic calendar from the Hijra (622 CE), the migration of Muhammad and his followers from Mecca to Medina to establish the first Muslim community (ummah). Similarly, the epoch for a Martian calendar should be defined by a culturally significant event. The question then becomes: what event will be (or is) culturally significant to Mars? The next "giant leap for mankind" has yet to be taken; in its absence, I originally chose the marsrev in which *Viking 1* began returning the first useful science from the surface of Mars, for it was this moment that mission controllers at JPL began operating according to the Martian diurnal cycle (sol). This choice had two problems: first, an American's selection of this event could be seen as nationalistic, although I know of no such objection having been raised; but secondly and much more problematic, Martians might desire to mark events prior to this one on their calendar, and if so, this would necessitate the inconvenience of counting marsrevs backward from the epoch. In 1999, Peter Kokh, a participant in the Martian Time Virtual Conference, suggested what came to be called the Telescopic Epoch, defined as the Martian northward equinox immediately preceding Galileo Galilei's first telescopic observations of Mars in 1610. That equinox occurred on 1609, March 11, and by this reckoning, the marsrev 220 begins on 2022, December 25. Coincidentally, this epoch also celebrates astronomer Johannes Kepler's publication of the *Astronomia nova,* which contained the results of his ten-year-long investigation of the motion of Mars. One of the most significant books in the history of astronomy, the *Astronomia nova* not only provided strong arguments for heliocentrism but presented the first two of his three laws of planetary motion: (1) that the planets move in elliptical orbits with the Sun at one focus; and (2) that the speed of a planet changes at each moment such that the time between two positions is always proportional to the area swept. Kepler's discoveries, in turn, led Sir Isaac Newton to formulate his laws of motion and his theory of gravitation, and thus the *Astronomia nova* is hailed as one of the most important works of the scientific revolution. We owe that revolution to the eccentric motion of Mars, which could not be adequately explained other than as elliptical. Certainly, Kepler's publication of the *Astronomia nova* and Galileo's telescopic observations were significant events to our modern technological civilization, and no doubt the future inhabitants of Mars will take a certain pride in their planet having beckoned humankind to embark on its scientific investigation of the nature of the universe.

The sol counts of the various surface stations and rovers on Mars, the Darian calendar date, Allison's Mars Sol Date, the solar longitude (L_S) of Mars, and Clancy's Martian epoch are not mutually exclusive[9], nor should they be viewed as competing with each other; rather, they complement each other by serving different timekeeping purposes. Accordingly, this almanac provides the sol count of ongoing and planned Mars surface operations[10], as well as the Mars Sol Date and the solar longitude of Mars, along with the Darian calendar date, in order to be useful to the widest readership. Also included in the date header for each sol is the Gregorian date and Universal Time Coordinated (UTC) when midnight occurs on the prime meridian on Mars (Airy Mean Time, or AMT). For instance, marsrev 220 begins with the following date header:

220 Sagittarius 01 **2022 December 25 11:11 UTC**

L_S 359.5° Curiosity Sol 3693 Perseverance Sol 657

MSD 52960

Following the date headers is a list of events for the given date in the present if any; for instance:

Dust storm season ends.

Northward equinox occurs at 22:30 AMT (December 26, 10:19 UTC).

This list of current events is followed by a description of past events that occurred on the sol, that is, the same date on the Darian calendar but in an earlier marsrev (the Martian time-slip: the past and present juxtaposed). For the first sol of marsrev 220:

On this sol in 160 (1910 February 18):

The motion picture *A Trip to Mars* was released. After inventing an anti-gravity powder, a man uses it on himself to propel himself to Mars.

—The Science Fiction Movie Database

On the Darian calendar, *A Trip to Mars* was released on 160 Sagittarius 01, which on the Gregorian calendar was 1910 February 18. The Martian anniversary for this event occurs on Sagittarius 01 every marsrev; however, Sagittarius 01 will not be February 18 for any instance other than the original event since the Earth year and the marsrev are of different durations.

An entirely terrestrial example of comparing dates on two calendars may be illuminating. Because the Gregorian calendar year is 365.2425 days and the Islamic calendar year is 354.37 days, the Islamic year 1443 began on 2021, August 10, according to the Gregorian

[9] Clancy's Martian epoch is a simple matter of subtracting 183 from the Darian marsrev count.

[10] For surface missions still en route to Mars at the time of publication, the sol count from the planned landing date is indicated; the actual landing date may differ.

calendar, while the year 1444 began on 2022 July 30. Later Islamic years may begin in April, or January, or October, eventually coming around to August again, yet on the Islamic calendar, the year always begins with the first day of the month Muharram.

Similarly, one's Martian birthsol is always the same date on the Darian calendar, but since it occurs once every 687 days, it occurs on different Gregorian dates from one marsrev to the next.

The Martian Almanac presents a number of drawings of the Red Planet made by astronomers over the centuries. It is particularly true of observations made with earlier and less powerful telescopes that the drawings can be difficult to reconcile with the albedo features, which became well-documented by the end of the 19th century. For each observation date, a computer-generated representation of the planet's aspect is presented to provide a side-by-side comparison. For example:

← 220 Kumbha 10 **2023 May 27 13:29 UTC →**

Ls 069.7° Curiosity Sol 3842 Perseverance Sol 806

MSD 53109

On this sol in 163
(1916 March 13):

Andrew E. Douglass observed and drew Mars. Syrtis Major and Nilosyrtis are identifiable.

According to the Pickering calendar, the Martian date was May 33.

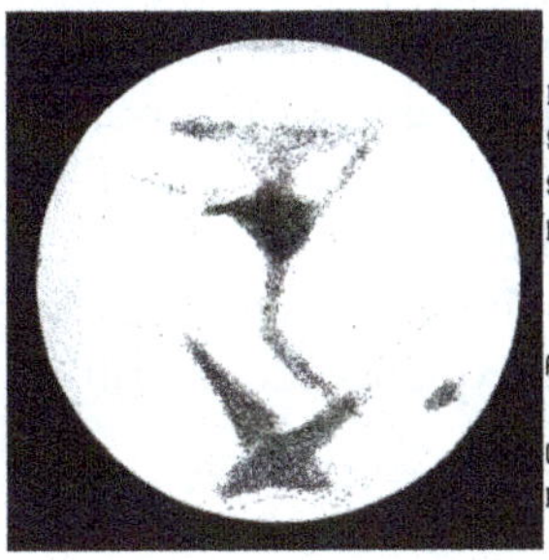

Douglass's drawing

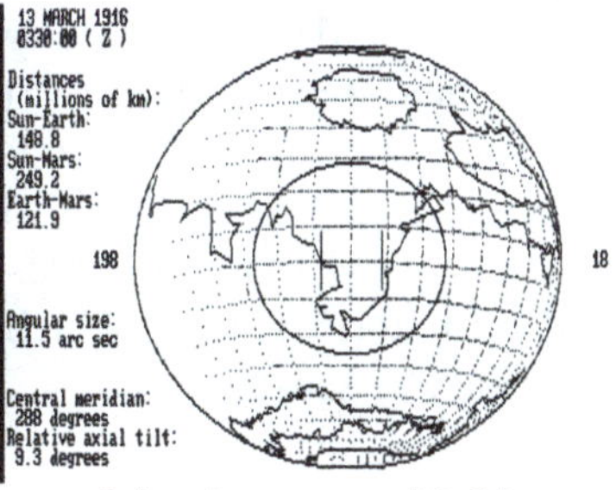

Mars's aspect, 23:00

The computer model generates a circle that varies in size in proportion to the distance between Earth and Mars on the date of the observation. Thus, a small circle represents an unfavorable opposition, as occurs when Mars is near aphelion (its farthest point from the Sun), around the Darian date of Khumba 13. The more detailed drawings of Mars tend to be rendered during favorable oppositions, when Mars is near perihelion (its closest point to the Sun), around the Darian date of Simha 13. For example:

220 Simha 15 **2024 May 09 21:10 UTC**
Ls 251.9° Curiosity Sol 4181 Perseverance Sol 1145
MSD 53448

On this sol in 100
(1798 August 27):

Johann Schröter drew a sketch of Mars. "Details just as well marked. Central meridian, 51°."

—Camille Flammarion, 1892

Aurorae Sinus and Erythraeum Mare are identifiable.

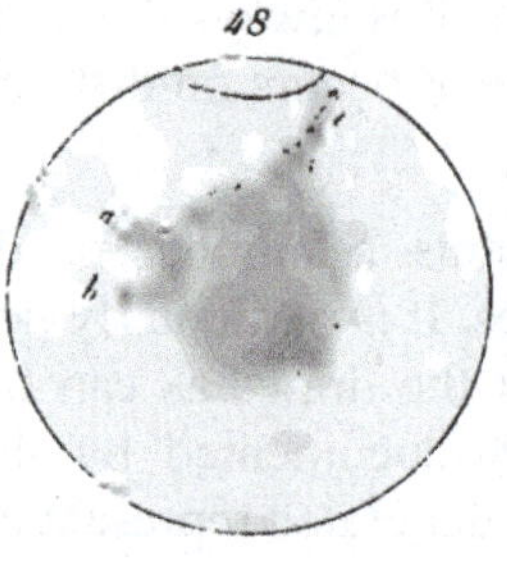

Schröter's sketch, 22:16

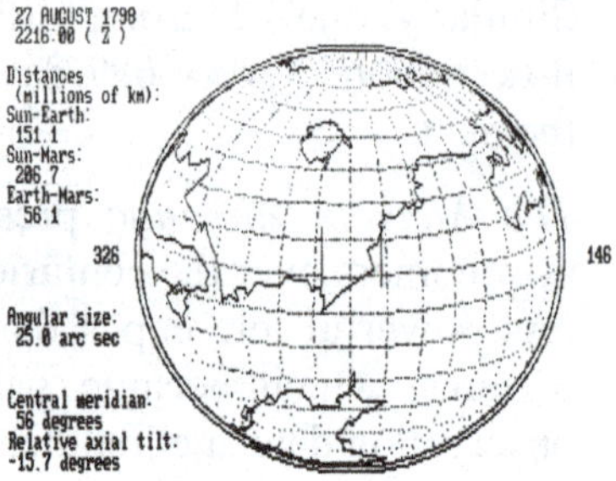

Mars's aspect, 22:16

It should be pointed out that during the early centuries, observers of Mars were situated in Earth's northern hemisphere; therefore, in the inverted images of the Red Planet presented in telescopes, the south was at the top and the north was at the bottom. Both the astronomers' drawings and the computer graphics reflect this convention. From this, it can be seen in the computer representations that during favorable oppositions, the Martian southern hemisphere is tilted toward Earth, and during unfavorable oppositions, the Martian northern hemisphere is tilted toward Earth. Thus, astronomers were generally able to discern more details of the southern hemisphere; however, favorable oppositions also occur at the height of dust storm season, which explains the dearth of detail in some of the sketches drawn during this season.

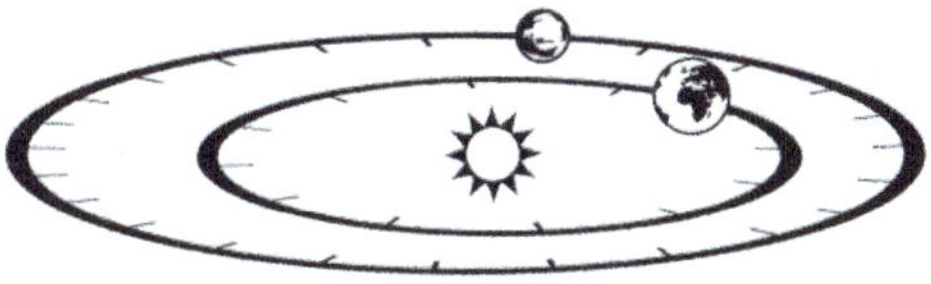

221 SAGITTARIUS

As seen from Mars, the Sun traverses the eastern sector of the constellation Sagittarius. The Babylonians identified Sagittarius as the god Nergal, a centaur-like creature firing an arrow from a bow. In turn, Nergal was identified with the planet Mars.

The median one-way communications delay during this month is 6 minutes 58 seconds.

221 Sagittarius 01 **2024 November 11 20:35 UTC**

Ls 359.7° Curiosity Sol 4361
MSD 53630 Perseverance Sol 1326

Dust storm season ends.

Northward equinox occurs at 12:49 AMT (November 12 09:46 UTC).

New Year's Sol. Feriae Marti, celebrated on Kalendis Martius on the Roman calendar (analogous to March 01 on the recalibrated Pickering calendar); dies natalis Martius (Mars's birthday), a feria also sacred to his mother Juno.

The first sol of the first quarter of Spring on the Aitken calendar; approximately the first sol of September on the Heinlein calendar.

On this sol in 171 (1932 September 15):

"The Electron Flame" by Jack Williamson appeared in *Wonder Stories Quarterly*.

Cover and interior art by Frank R. Paul.

On this sol in 210 (2004 March 04):

MER-B Sol 40: *Opportunity* finished a set of overnight alpha particle X-ray spectrometer measurements at Last Chance and completed a morning set of panoramic camera and miniature thermal emission spectrometer remote sensing observations. At 11:30 local solar time, engineers retracted the alpha particle X-ray spectrometer off the target, took a final set of 24 microscopic images, and stowed the arm for driving.

Opportunity then scored another first by successfully using visual odometry to navigate autonomously on Mars. During a drive along the crater wall, the vehicle properly identified wheel slippage on the steep slope of the crater wall using features in the navigation camera imagery. This effectively provided a mid-course correction that landed the science and engineering team exactly at the target location where they want *Opportunity* to do work using the instruments on the rover arm on Sol 41.

The plan for Sol 41 will be to take microscopic images of an area dubbed Wave Ripple in the Last Chance area, followed by a traverse to Slick Rock in the Berry Bowl area.

On this sol in 210 (2004 March 05):

MER-A Sol 61: After more than a week of camping and field work at Middle Ground, NASA's *Spirit* took a few last pictures from there then drove onward to the northeast. In the morning, *Spirit's* panoramic camera took the final frames needed for the camera team to assemble a full-circle color panorama after all the data reaches Earth.

In the early afternoon, *Spirit* backed up 0.5 meter (20 inches), then edged forward 0.29 meters (11 inches) to sidestep a rock called Ingrid. Then the rover advanced 28.5 meters (94 feet) toward its crater-rim destination. The drive took 45 minutes. From the new location, *Spirit* took forward-looking pictures for use in future drive planning. It also observed the ground and the sky with its miniature thermal emission spectrometer.

—NASA/JPL

221 Sagittarius 02 **2024 November 12 21:15 UTC**
Ls 000.2° Curiosity Sol 4362
MSD 53631 Perseverance Sol 1327

On this sol in 152 (1895 February 03):

Andrew E. Douglass observed the canals Gihon, Typhon, Hiddekel, Orontes, Deuteronilus, Phison, Oceanus, Protonilus, Casuentus, and Acalandrus.

According to the Douglass calendar, the Martian date was March 17.

On this sol in 210 (2004 March 05):

MER-B Sol 41: NASA's *Opportunity* inspected a rock target called Wave Ripple with tools on its arm, then drove to a new target. The new target, Flat Rock, is in the Slick Rock area near the south end of the outcrop that the rover has been examining for weeks.

Although the rover wheels slip some in the local soil and the drive traversed a slope of 10 to 11 percent, *Opportunity* and engineers at JPL navigated the trip so well that a planned final approach to the target on Sol 42 (210 Sagittarius 03) could be cancelled. The target is within the work volume of *Opportunity's* robotic arm. The drive was done in a series of one-meter (3.3-foot) segments making up a U-shaped path to the south and west. Each segment included a correction for slippage.

Before starting the drive, *Opportunity* used its microscope for 50 images of Wave Ripple, and examined the composition of the rock with its alpha proton X-ray spectrometer and its Mössbauer spectrometer.

On this sol in 210 (2004 March 06):

MER-A Sol 62: NASA's *Spirit* advanced about one-fifth of the remaining distance between where it began the sol and its mid-term destination, the rim of the crater nicknamed Bonneville. In the afternoon, *Spirit* took images and infrared readings of the area right in front of its stopping place to support the following morning's close-up inspection of that new location with instruments on the rover's robotic arm.

Spirit drove 26.15 meters (85.8 feet) on Sol 62, bringing its odometer total to 250.71 meters (822.5 feet). Some of the drive maneuvered around obstacles. The net gain in the northeasterly direction toward the crater rim was 22 meters (72 feet), and that destination

was estimated to be about 88 meters (289 feet) away from *Spirit's* new location. The miniature thermal emission spectrometer was used for ground and sky observations both before and after the drive.

—NASA/JPL

221 Sagittarius 03 **2024 November 13 21:54 UTC**

Ls 000.7° Curiosity Sol 4363

MSD 53632 Perseverance Sol 1328

On this sol in 152 (1895 February 04):

Andrew E. Douglass observed the canals Nilosyrtis, Typhon, Orontes, Phison, Protonilus, Deuteronilus, Casuentus, Acalandrus, Oceanus, Tedanius, and Cephissus.

According to the Douglass calendar, the Martian date was March 18.

On this sol in 159 (1908 April 04):

MARS IN VISIONS AND IN ACTUALITY

Present State of Reliable Knowledge About the Planet Is Summed Up by Prof. Poor.

All of us acquired at school an exact knowledge of that portion of the contemporary astronomy which professed to describe and chart the motions, orbits, and characteristics of the sun and its attendant planets. So did the schoolboys of Nineveh and Alexandria, no doubt, acquire a like exact knowledge of their contemporary astronomy. Astronomies are, however, like the schedules of railway trains, subject to change without notice. And especially the unqualified statements inserted in elementary textbooks on astronomy—supposedly adapted to the severe simplicity of the youthful mind—these unqualified statements frequently need revision. Therefore all conscientious persons whose school days are over ought to read a brand-new astronomy book at least once a decade.

For that purpose and for the present decade such conscientious persons may be confidently recommended to "The Solar System; A Study of Recent Observations" written by Charles Lane Poor, Professor of Astronomy in Columbia University, and published by G. P. Putnam's Sons at their Knickerbocker Press. The work in question is compact, readable, and instructive in just the points where one's schoolboy ideas need readjustment.

The task of readjustment at large, agreeable as is the process of passing on to those who still sit in darkness, an enlightenment which but yesterday was ignorance as black as theirs—this task is not for the reviewer. Prof. Poor shall be permitted to do the work himself. But insomuch as another professor, Percival Lowell, has of late been seeing things at night from Flagstaff, Arizona, and insomuch as, on the strength of these visions, the planet Mars has entered upon one of those periods of "news interest" to which all planets and all men are subject, it seems only fair to give the reader the immediate benefit of our author's admirable review of the present state of knowledge—including all that can be argued from Prof. Lowell's visions—with regard to our nearest heavenly neighbor but one.

The final summing up shall be put first. "From all the conflicting data one conclusion may safely be drawn, and this is that very little is actually known in regard to the conditions existing on Mars." Prof. Lowell, as we have all read, sees the ruddy planet covered with an

intricate network of fine straight lines, which he has convinced himself have real Martian objectivity and a purpose. He argues the purpose from the straightness and from another thing, which he has convinced himself is true of the lines—their origin in and radiation from the polar regions. According to him, the lines he sees are the cultivated belts which border a network of canals designed to irrigate the whole arid surface of the planet from the melting snows of the polar ice caps. Hence of course Martians, intelligent beings, irrigators—perhaps the Martian Universal Water and Power Company, Ltd., a trust by the side of which the Standard Oil is child's play.

Prof. Poor sets out Mr. Lowell's observations and inferences with entire fairness. But he calls attention to several facts which seem to weaken Lowell's case. For instance, other astronomers with almost equally good facilities fail to see. or see quite differently, that marvelous network of lines which as optical phenomena art, if they exist, "at the very limit of our vision; on the borderland where perception ends and illusion begins." For "the visible disk is minute, and is seen through many miles of the shifting, trembling atmosphere. With the magnifying powers ordinarily used Mars would appear somewhat the size of a silver quarter at the bottom of a stream three or four feet in depth.

Currents and disturbances in the water render objects at the bottom hazy and indistinct; only at moments of perfect quiet can the lettering of the coin be seen. So some astronomers see the straight lines as irregular shadings merely and refer the apparent geometrical form to an ocular effort to arrange these irregular shadings.

Even supposing the lines straight; Pro. Poor fails to see evidence of the intelligent irrigator. Mars is a planet of uneven surface. Would an intelligent irrigator build his canals straight through hills and valleys in defiance of the law of gravity? And how could the water flow from the pole at one season and from the opposite pole at the other season? Also there is the difficulty of the Martian atmosphere. Prof. Poor finds Prof. Lowell playing strange tricks with that. "On the one hand, he explains the presence of and the artificiality of the canals by the scarcity of water upon the planet, by the necessity of husbanding every drop of the precious fluid; on the other he accounts for the temperature necessary for the existence of free water by assuming an atmosphere laden with water vapor. He conjures up a dry, parched desert in which sandstorms abound, covered over with a moist, saturated atmospheric blanket!"

Without such an atmosphere the "average temperature of the planet" must be "much lower than that of the earth, and is probably below the freezing point of water. The theoretical temperature is minus 22 degrees Fahrenheit." The spectroscope fails to establish the existence of water at all, which is argued from the waxing and waning with the seasons of those lighter areas which have been supposed (by analogy of the earth) to represent the polar ice caps.

In short, Prof. Lowell's Martians and their canals are at best a theory, not a condition. Prof. Poor declines to accept them as objective realities. Once more: "But one conclusion may safely be drawn; that very little is known in regard to conditions existing on Mars."

—*The New York Times,* page 18

On this sol in 210 (2004 March 06):

MER-B Sol 42: NASA's *Opportunity* attempted to grind a shallow hole into a target called Flat Rock. However, the operation of the rover's rock abrasion tool produced almost no discernable impression on the rock. All indications are that the tool is healthy. Controllers plan to run some diagnostic tests during Sol 43 (210 Sagittarius 04) to aid with tuning parameters for a second grinding attempt on the target on Sol 44 (210 Sagittarius 05).

Opportunity observed the Sun with its panoramic camera on Sol 42 as a practice run for future imaging of Mars's moon Phobos passing in front of the Sun. The alpha particle X-ray spectrometer was placed against Flat Rock for an overnight reading to identify the chemical elements present.

On this sol in 210 (2004 March 07):

MER-B Sol 63: The panoramic camera on Mars Exploration Rover Spirit recorded a streak which is believed to have been caused by a meteor shower associated with comet 114P/Wiseman-Skiff occurred over Mars.

—NASA/JPL

221 Sagittarius 04 **2024 November 14 22:34 UTC**

Ls 001.2° Curiosity Sol 4364
MSD 53633 Perseverance Sol 1329

On this sol in 171 (1930 November 01):

"Missionaries from the Sky" by Stanton A. Coblentz appeared in *Amazing Stories* (date approximate). Aliens living in a perfect utopia offer to come to Earth's aid and end hunger, violence, social inequality, and all humankind's other ills – at a very small cost.

Interior art by Frank R. Paul.

"A Message from Mars" by Derek Ironside appeared in *Weird Tales* (date approximate). Professor Bullivant, who is something of an academic thug, has built a space ship that will take him to Mars in about a month. Builivant, who has taken along a band of roughnecks and a small armory. is obviously heading for trouble.

—Edward Franklin Bleiler, 1990

On this sol in 210 (2004 March 07):

MER-B Sol 43: Using a combination of microscopic images, hazard avoidance camera images, and rock abrasion tool tests on Sol 43, *Opportunity's* engineering team discovered that the grind motor of the rock abrasion tool on Mars stalled prematurely during what's called the "seek/scan" phase when the rock abrasion tool instrument searches for the rock face. This resulted in no contact during the actual grind activity on Sol 42 (210 Sagittarius 03). The most likely causes of the stall are dust and dirt accumulations and temperature variations on the instrument. The tests also confirmed that engineers can safely increase the motor voltage on the instrument to prevent a future stall.

In the process of conducting the rock abrasion tool activities, *Opportunity* placed the Mössbauer spectrometer on Mojo 2, a target on Flat Rock.

The remainder of the sol included preparations for the future target in this area, dubbed Berry Bowl, taking pictures with the panoramic and navigation cameras. *Opportunity* also took ground and atmospheric measurements with the miniature thermal emission spectrometer.

A transit of Phobos occurred over the area.

On this sol in 210 (2004 March 08):

MER-A Sol 64: *Spirit* began the morning with an 18-meter (59 feet) direct drive that safely maneuvered the rover through a field of rocks. *Spirit* then traversed 11 more meters (35 feet) using autonomous navigation and at 11:30 a.m. completed the drive. *Spirit* had some difficulty finding a way around an obstacle during the last portion of the commanded drive. That resulted in some repeated forward and backward maneuvering which left an interesting trench for scientists to have the rover peer into.

Spirit completed another 29 meters (94 feet) of its drive toward the rim of Bonneville crater, bringing its total odometry to 314 meters (1,030 feet) - 14 meters (45.9 feet) past the minimum mission success criterion.

Spirit was climbing up a very steep part of Bonneville now, and ended this sol's drive tilted at a forward pitch of about 15°.

—NASA/JPL

221 Sagittarius 05 **2024 November 15 23:13 UTC**

Ls 001.8° Curiosity Sol 4365
MSD 53634 Perseverance Sol 1330

Tianwen-1 is scheduled to begin its third marsrev of orbital operations.

On this sol in 210 (2004 March 08):

MER-B Sol 44: *Opportunity's* the rock abrasion tool ground a 3.1 millimeter-deep (just over one-tenth of an inch) hole in the Mojo 2 target on Flatrock. Yestersol, diagnostic testing determined a voltage adjustment was necessary to overcome some mechanism "stickiness" in the routine during which the rock abrasion tool finds the highest point in the target area.

The routine worked perfectly on this grind with the new voltage setting. After one hour and five minutes of successful grinding, the rock abrasion tool grind motor stalled, probably while grinding into one of the spherules also known as "blueberries." These objects are known to obstruct the grinding tool and cause it to terminate its sequence.

Late in the Martian morning, the Mössbauer spectrometer was placed on the hole, followed later by the alpha particle X-ray spectrometer. The miniature thermal emission spectrometer took two long atmospheric measurements. The panoramic camera was busy taking images of the hole as well as surrounding target areas.

On this sol in 210 (2004 March 09):

MER-A Sol 65: *Spirit* spent the sol analyzing soil targets with the alpha particle x-ray spectrometer, Mössbauer spectrometer and microscopic imager before stowing its arm, doing some remote sensing of the trench dubbed Serendipity Trench, and then finally setting off for the longest directed drive to date. That drive was 27 meters (88.6 feet) toward the edge of Bonneville crater.

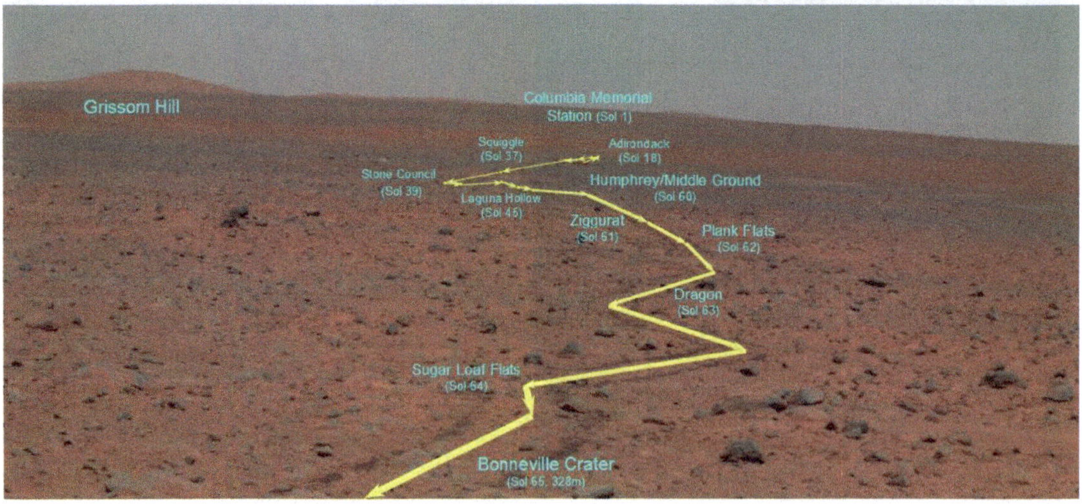

Looking back from Bonneville crater to the landing site.

Spirit then attempted to use auto navigation to reach a target that was an additional 6 meters (19.7 feet) away. Sensitive obstacle avoidance software prevented *Spirit* from reaching the destination, and like yestersol, the rover completed several drives forward and back. Those drives resulted in a final odometer reading of 40.7 meters (133.5 feet) for the sol, even though the total straight-line distance traveled was 30 meters (98.4 feet).

The 30-meter (98.4 feet) drive put *Spirit* close enough to Bonneville's edge to take images with the navigation cameras that reveal the opposite rim of the crater.

—NASA/JPL

221 Sagittarius 06 **2024 November 16 23:53 UTC**

Ls 002.3° Curiosity Sol 4366

MSD 53635 Perseverance Sol 1331

Earth departure opportunity to Mars via a Type I trajectory.

Earth departure opportunity to Mars via a Type II trajectory.

Earth arrival opportunity from Mars via a Type II trajectory

On this sol in 155 (1900 September 29):

Arthur Strawn was born; American screenwriter; wrote the screenplay for the motion picture *Flight to Mars* (1951).

On this sol in 186 (1959 January 19):

Quatermass and the Pit, Episode 5, "The Wild Hunt" air date. A brain scanner reveals alien pictures in Barbara Judd's mind of "the Wild Hunt"—the frenetic purging of the Martian hives. In the pit, a technician is killed as the capsule starts to glow.

—International Movie Database

Quatermass theorizes that the human race was genetically engineered by insectoid Martians from a more primitive anthropoid species to house the intelligence of their colonists, while the original Martian race died out. Fullalove responds, "So, as far as anybody is, we're the Martians now."

On this sol in 210 (2004 March 09):

MER-B Sol 45: *Opportunity* is halfway through its primary 90-sol surface mission.

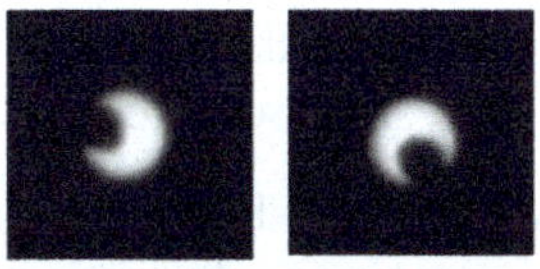

Opportunity snapped two images of Mars's moon Phobos as it passed across the face of the sun.

Opportunity used the rock abrasion tool brush to sweep off the dirt in and around the hole at Mojo 2 in the Flat Rock area. *Opportunity* then took five microscopic images of the freshly brushed Mojo 2.

The miniature thermal emission spectrometer took measurements at three locations on the surface of Mars, and then pointed upwards to observe the atmosphere in four different directions. The panoramic camera was also busy taking images of the magnets around the rock abrasion tool area, Mojo 2 post brushing, and a new area called Slick Rock.

On this sol in 210 (2004 March 10):

MER-A Sol 66: After a 21-meter (68.9 feet) drive and an elevation gain of 1.5 meters (4.9 feet), *Spirit* is at the edge of Bonneville crater.

Spirit began Sol 66 by taking a look back at the lander with the panoramic camera, and then analyzing selected ground targets. At about 11:30 a.m. Gusev time, *Spirit* began a 30-minute directed drive of 16 meters (52.5 feet). It then turned right for a final auto-navigation drive that was intended for 6 meters (19.7 feet), but resulted in about 4 meters (13.1 feet). That drive put Spirit in perfect position to look inside Bonneville crater and send back stunning images from the navigation camera.

The camera mast was then positioned for a nighttime sky observation of the constellation Orion.

—NASA/JPL

221 Sagittarius 07 **2024 November 18 00:32 UTC**
Ls 002.8° Curiosity Sol 4367
MSD 53636 Perseverance Sol 1331

Predicted meteor shower associated with parent body 13P/Olbers: Martian right ascension 27.5°, declination -44.7°. Impacting at 27 km/sec on the southern hemisphere of Mars well into the night side.

On this sol in 151 (1893 March 22):

In *Journey to Mars* (1894) by Gustavus W. Pope, merchant seaman Tom Hallyard, first mate of the *White Gull,* delivered to the *Albatross's* physician in New York the journal of Lieutenant Frederick Hamilton documenting his journey to Mars.

On this sol in 210 (2004 March 10):

MER-B Sol 46: *Opportunity* performed a series of activities including microscopic imaging of the berries and placing the Mössbauer spectrometer on the berries to analyze their chemical composition. The miniature thermal emission spectrometer later made multiple atmospheric observations. After a short nap to conserve energy, *Opportunity* awoke in the afternoon to perform some additional remote sensing observations and to transmit data to Earth via the *Mars Odyssey* orbiter.

A transit of Phobos occurred over the area.

Later in the evening local solar time, Opportunity collected data with its alpha particle X-ray spectrometer at two locations.

The plan for Sol 47 is to continue analyzing the blueberries and the Berry Bowl. By early next week, Opportunity will drive to a new area dubbed Shoemaker's Patio.

On this sol in 210 (2004 March 11):

MER-A Sol 67: Before beginning the Sol 67 drive, *Spirit* completed an overnight alpha particle x-ray spectrometer integration and a couple of small panoramic camera shots of its surroundings.

Then the rover traveled 13 meters (42.7 feet) in a direct drive around some obstacles followed by a 1.4-meter (4.6 feet) automatic navigation drive. *Spirit* spent the afternoon using the mini thermal emission spectrometer to look at targets that will be analyzed more fully on Sol 67, and then driving 0.9 meters (3 feet) forward to be able to access that area with the arm tomorrow.

Before the sol was over, *Spirit* looked at the Sun for an attitude update and then took front hazard avoidance camera images of the arm work volume, and a small navigation panorama of the crater. The rover also completed some mini thermal emission spectrometer analysis of the far side of the crater and finished the day with some panoramic camera images of the sunset.

Spirit put a total of 24.8 meters (81.4 feet) on the odometer this sol, partly due to some back and forth maneuvering it had to do to ensure a safe path. The final location has proven to be just right, and *Spirit* will stay put for a couple sols while it continues to investigate Bonneville crater.

—NASA/JPL

---

221 Sagittarius 08 **2024 November 19 01:12 UTC**

Ls 003.3° Curiosity Sol 4368
MSD 53637 Perseverance Sol 1333

On this sol in 152 (1895 February 09):

Fyodor Aleksandrovich Otsep (Фёдор Александрович Оцеп) was born; Russian/Soviet film director, screenwriter, and actor; contributing writer to the screenplay for *Aelita: Queen of Mars (Аэлита – королева Марса)* (1924).

Max Valier was born; Austrian rocketry pioneer; author of *A Daring Trip to Mars* (1931).

On this sol in 176 (1940 April 01):

"Murder from Mars" by Richard Wilson appeared in *Astonishing Stories* (date approximate). Ray Carver, top-notch Earth detective, looks for the first interplanetary killer and finds a wife.

Interior art by Joseph Lasker.

"The Treasure of Ptakuth" by Leigh Brackett appeared in *Astounding Science-Fiction* (date approximate). The treasure was invaluable, inestimable—like the product of infinity times zero. For, in a way, it was both—zero and infinity, too.

Interior art by Hubert Rogers.

"Stepsons of Mars" by C. M. Kornbluth, Richard Wilson, and Dirk Wylie also appeared in *Astonishing Stories* (date approximate). Guns are not enough, when the enemy is too strong. But if you can find out their weak point, and if you have one man who has the weapon to be used against it….

Interior art by Hannes Bok.

On this sol in 186 (1959 January 22):

The Unforseen, Season 1 Episode 16, "The Monsters" airdate. Two scientists attempt a rocket trip that could make them the first Martians to set foot on the planet Earth.

On this sol in 210 (2004 March 11):

MER-B Sol 47: *Opportunity* performed a series of activities including microscopic imaging of the berries and placing the Mössbauer spectrometer on the berries to analyze their chemical composition. The miniature thermal emission spectrometer later made multiple atmospheric observations. After a short nap to conserve energy, *Opportunity* awoke in the afternoon to perform some additional remote sensing observations and to transmit data to Earth via the *Mars Odyssey* orbiter.

Opportunity finished remnants of activities from the past sol's research at Berry Bowl. The sol started with the hazard avoidance camera taking a picture of the Berry Bowl area as a context picture. The miniature thermal emission spectrometer then performed some sky stares of the atmosphere. At 11:30, the robotic arm started moving. It picked up the alpha particle X-ray spectrometer and moved to a new location nearby, then switched to the Mössbauer spectrometer. Both spectrometers are searching for clues about the chemical composition of the mysterious blueberries.

Later, *Opportunity* took panoramic camera images of the suite magnet on the rover itself, which is collecting atmospheric dust samples to understand why the Martian dust is so magnetic. The panoramic camera also took images of a target dubbed Fool's Silver, which contains an interesting angular feature in the outcrop.

After all the morning's hard work, *Opportunity* took a short siesta to rest and recharge. *Opportunity* reawakened a few hours later to take more images of the atmosphere with the miniature thermal emission spectrometer and panoramic camera. Those were taken in the same locations as the morning measurements to compare the atmospheric data throughout the sol.

At 15:40, *Opportunity* took about a dozen images of the Sun to catch the transit by the Martian moon, Phobos.

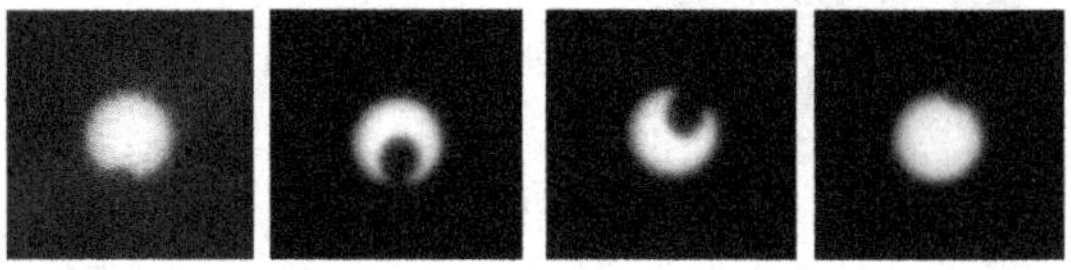

On this sol in 210 (2004 March 12):

MER-A Sol 68: *Spirit* spent all of Sol 68 at the Bonneville crater location. It began the morning operating the panoramic camera to acquire the first images of what will be a 360-degree shot of Bonneville's rim and basin, and the Columbia Hills to the southeast. The hills were named to honor the crew of Space Shuttle *Columbia's* ill-fated STS-107 mission.

Spirit also moved the instrument deployment device, or rover arm, into position to acquire panoramic camera images of the magnets on the rock abrasion tool. It then placed the Mössbauer spectrometer on soil for a short integration after taking five microscopic imager images.

Around 13:35, one of Mars's moons, Deimos, passed in front of the sun. Scientists and rover controllers took this opportunity to image the moon's transit with the panoramic camera before completing mini thermal emission spectrometer observations of the crater interior.

—NASA/JPL

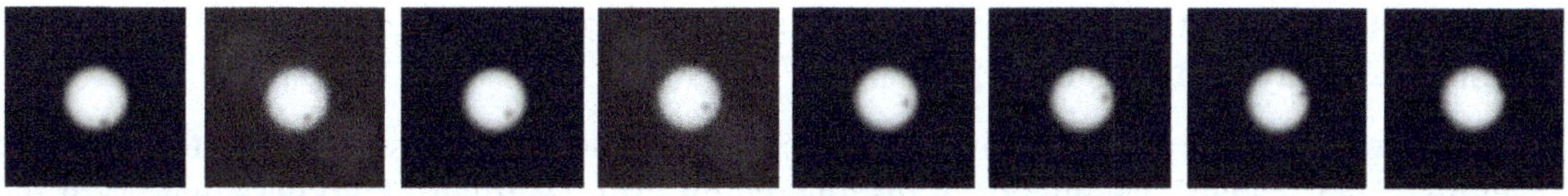

221 Sagittarius 09 **2024 November 20 01:52 UTC**

Ls 003.8° Curiosity Sol 4369
MSD 53638 Perseverance Sol 1334

The first sol of April, marsrev 3679 on the Levitt calendar.

On this sol in 152 (1895 February 10):

NEWS FROM MARS.

Of all the planets Mars is the one which most closely resembles the earth. Astronomers are generally of the opinion that Mars is inhabited. The argument in favor of the existence of intelligent beings in Mars is plausible, but not entirely conclusive. Beyond a doubt snow falls in the polar regions of Mars. Hence rain must also fall in the warmer regions. If there is rain, there may also be beer, for water is the principal constituent of beer; and if there is beer, there are, of course, intelligent beings to drink it. This argument was considered by Sir Isaac Newton to be conclusive, but there is more than one eminent astronomer at the present day who regards the argument as plausible rather than conclusive. Mars has latterly been closely studied by astronomers in order to ascertain, if possible, the nature and origin of certain lines on the surface of the planet, which are commonly thought to be canals.

Water, presenting as it does a smooth surface, reflects light. It has therefore been assumed that the bright lines which stretch from one part of Mars to another and seem to connect broader surfaces of light are canals. There are, however, two difficulties in the way of the acceptance of this theory.

A so-called canal at least a thousand miles in length will often make its appearance in a single night. Now, the Martians may be very clever and very industrious, but it is doubtful if they can construct a ship canal a thousand miles long in twenty-four hours. Then, again, these canals are not permanent, and it often happens that one of them disappears entirely after having been in existence for a. few days or months. What is still more curious is the fact that these canals often appear to be double.

The first time that an astronomer reported that he had seen two parallel canals running side by side his brother astronomers agreed that it was very sad, and that he had better confine himself exclusively to tea. But the double canals have since been seen so frequently and by astronomers of such proved sobriety, so far as alcohol is concerned, that it is no longer possible to doubt their existence. Only a few days ago one of the leading German astronomers reported that the Erie Canal, which is one of the longest of the Martian canals, had been duplicated in the course of a single night. This report, accompanied as it was by the certificate of a physician, who asserted that the astronomer had consumed but three glasses of beer on the night in question, has been generally accepted among astronomers as trustworthy. If, then, the canals are really artificial waterways, we must believe that the Martians can dig a thousand miles of canal in a few hours, and then that the digging of such a canal is followed by such a growth of the local traffic that a second canal, parallel to the first, is usually required. Now, a habitual astronomer can believe a good deal, but these alleged Martian feats of canal digging are so tremendous that astronomers are beginning to doubt if the so-called canals are really anything of the sort, and several new explanations of them have lately been suggested.

It is not worth while to mention these various suggestions, for it so happens that the true nature of the "canals" has just been discovered by Prof. [William] Harkness, who is at the head of the United States Naval Observatory in Washington. This observatory contains a very powerful telescope, which was originally intended to be used in searching for parallels of latitude and meridians of longitude and reporting their location to the commanders of American men-of-war. Since all the parallels and meridians have long since been discovered and their location accurately described, the chief astronomer of the Naval Observatory has turned his attention and his telescope to the heavens and has made a number of important discoveries. Prof. Harkness's predecessor was under the impression that the planet Mars was under the exclusive jurisdiction of the War Department, and hence never turned his telescope in the direction of Mars. Prof. Harkness, however, knew better, and one of the first things which he did after he was formally invested with the telescope was to point it at Mars and to examine the "canals" with the greatest care. In a very short time he discovered that what previous astronomers had supposed to be canals, merely because they were straight lines of reflected light, were nothing more than clothes lines filled with the week's wash of the neighboring inhabitants.

Nothing reflects light better than white linen. It we assume that Mars is densely inhabited and that the inhabitants wear a dress somewhat resembling the Roman toga, the extent of Martian clothes lines is fully explained. Prof. Harkness claims that he has not only recognized the pattern of several of the Martian garments, but that he has also perceived in several instances the names of the owners embroidered in the corners of togas and the waistbands of skirts. He is of the opinion that the clothes lines of Mars are under the control of the Government, which wisely stretches them in straight lines through the midst of the most thickly populated districts, and that whenever an unusually large "wash" is ready for the line the Government puts up extra lines to meet the emergency and thus brings about the phenomenon which has hitherto been mistaken for a sudden duplication of imaginary canals.

The importance of this discovery will be recognized by all astronomers. The canal theory had become almost untenable, and astronomers clung to it because they had no other theory to take its place. The clothes-line theory is, however, free from all possible objections, and Prof. Harkness can proudly claim that he has virtually proved it to be true.

—*The New York Times,* page 4

On this sol in 210 (2004 March 12):

MER-B Sol 48: *Opportunity* once again shut down for a nap and woke up at 4:53, Sol 48, for a tool change and a communications session with the *Mars Odyssey* orbiter. While the rover was awake for the Odyssey pass, the rover heated up the robotic arm, which had chilled to almost -80 degrees Celsius (-112 degrees Fahrenheit). The motors cannot move at that frigid temperature, so the rover arm heated for 32 minutes to surpass the operational temperature of 40 degrees Celsius (104 degrees Fahrenheit). As the rover arm quickly cooled, the heat lasted long enough (5 minutes) for the arm to twist its wrist and change instruments from the Mössbauer spectrometer back to the alpha particle X-ray spectrometer.

Opportunity finished inspecting the Berry Bowl site and drove 10 meters (33 feet) toward a new target. The rover used all four tools on its arm during the morning, ending with a brushing by the rock abrasion tool, then post-brushing examinations with the microscope and alpha particle X-ray spectrometer. This closed out three sols of work at Berry Bowl to compare the composition of targets with and without groups of the BB-sized spherules believed to have formed while the local environment was wet.

Opportunity then stowed its arm and drove toward an area dubbed Shoemaker's Patio at the southwestern end of the outcrop the rover has been studying since it arrived on Mars. This informal name pays tribute to the late geologist Dr. Eugene Shoemaker of the U.S. Geological Survey. *Opportunity's* more specific target is a rock called Shark's Tooth at the near edge of the patio. The drive did not quite put that target within reach of the robot arm. Activities of the sol also included atmospheric observations with the panoramic camera and miniature thermal emission spectrometer, plus post-drive imaging with the navigation camera.

A transit of Phobos occurred over the area.

On this sol in 210 (2004 March 13):

MER-A Sol 69: This panoramic image mosaic from the Mars Exploration Rover *Spirit* panoramic camera, shows the rover's destination toward the hills nicknamed the Columbia Hills, on the right. The rover's heatshield can be seen on the left as a tiny bright dot in the distance, just under the horizon. Dark drift material can be seen in the image center. The rover is currently positioned outside the view of this image, on the right. This image was taken on Sols 68 and 69 of *Spirit's* mission (210 Sagittarius 08 and 09) from the location the rover first reached on the western rim of the crater. The image is in approximate true color, based on a scaling of data from the red, green, and blue (750 nanometers, 530 nanometers, and 480 nanometers) filters.

Spirit finished shooting frames with its panoramic camera for a full 360-degree color view of the surroundings visible from the crater rim where the rover is perched. Once the panorama frames are transmitted to Earth, scientists will use them and information from *Spirit's* miniature thermal emission spectrometer to assess the structures and composition of the crater interior and other surfaces in view.

In the afternoon, *Spirit* added a set of observations with the miniature thermal emission spectrometer in coordination with overhead passage of NASA's *Mars Global Surveyor* orbiter, which carried a similar spectrometer looking down. Events of the sol also included two camera sessions requested by engineers. The first was to get high-resolution images of *Spirit's* heat shield on the northern rim of Bonneville. The other was to photograph wheel tracks to help rover mobility specialists assess slippage.

Spirit did not move from its vantage point on the south rim of the crater nicknamed Bonneville. An extra downlink session was added via relay by NASA's *Mars Odyssey* orbiter to accelerate getting the panoramic imaging data to Earth.

—NASA/JPL

221 Sagittarius 10 | **2024 November 21 02:31 UTC**

Ls 004.3° | Curiosity Sol 4370

MSD 53639 | Perseverance Sol 1335

On this sol in 194 (1974 February 10):

Mars 4 failed to enter Mars orbit and passed Mars at a distance of 1,844 kilometers (1,146 miles), returning 12 images mostly in the Argyre quadrangle (MC-27) from its photo-television camera, all through the red filter, eight of which were of usable quality. It also returned two panoramas from the optical-mechanical cameras.

Panorama 2

Image 4-1-7-B

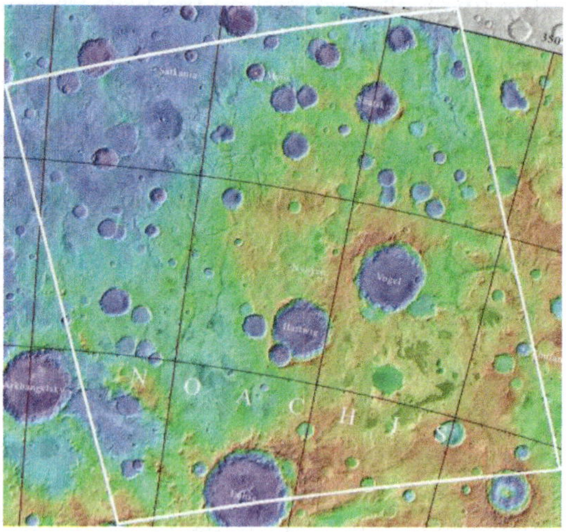

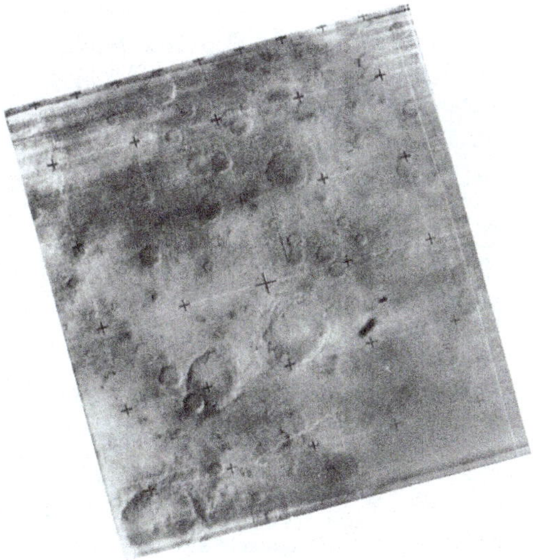

Image 4-1-8-B

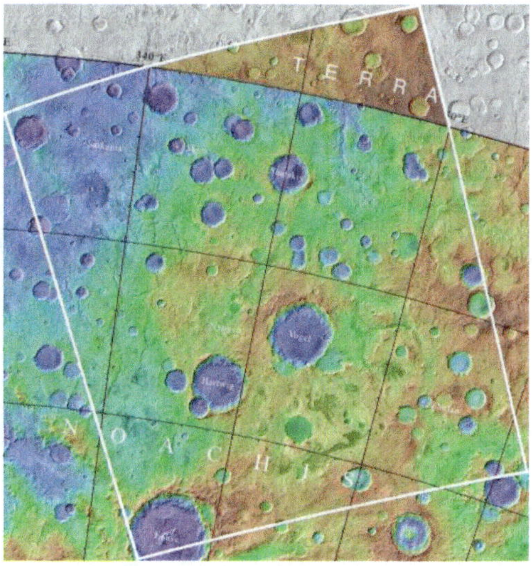

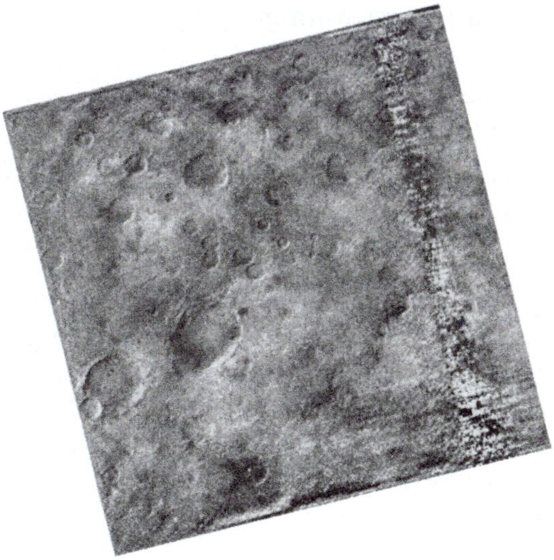

Image 4-1-9-B

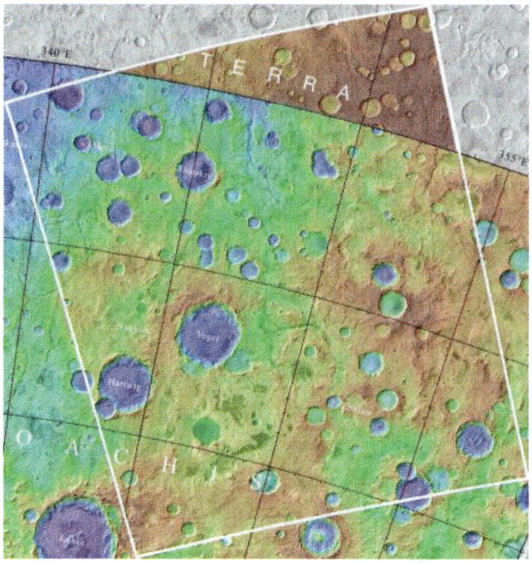

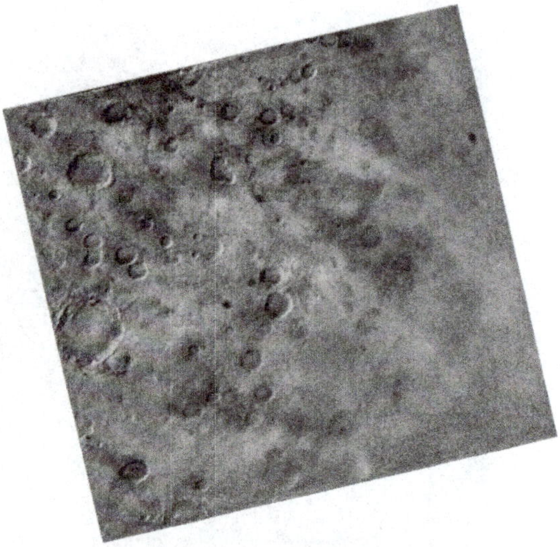
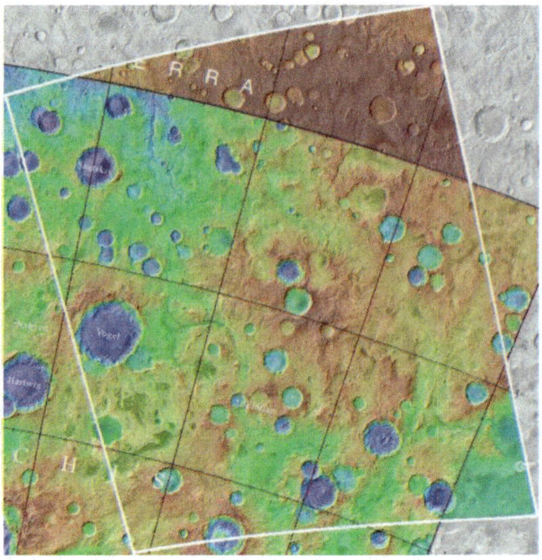

Image 4-1-10-B

See *The Martian Almanac 220, Volume 1* for the earlier images from the flyby.

On this sol in 210 (2004 March 14):

MER-A Sol 70: *Spirit* began what would be a very busy sol by analyzing a soil target dubbed Gobi 1 with the Mössbauer spectrometer. This was the first of 43 observations that *Spirit* would complete on Sol 70, breaking the previous observation record of 31 observations in one sol.

After the successful Mössbauer integration, *Spirit* took panoramic camera images of the sky. Then the miniature thermal emission spectrometer analyzed rock and soil targets. Following this, *Spirit* turned its panoramic camera and miniature thermal emission spectrometer to a range surface and atmospheric observations. Before the sol ended, *Spirit* also took microscopic images of the Mössbauer footprint left on the soil target and switched instruments to the alpha particle X-ray spectrometer for a long integration starting early on Sol 71 (210 Sagittarius 11) at the Gobi 1 location.

—NASA/JPL

221 Sagittarius 11 **2024 November 22 03:11 UTC**

Ls 004.8° Curiosity Sol 4371

MSD 53640 Perseverance Sol 1336

On this sol in 141 (1874 June 05):

Enrico Novelli was born; Italian director, writer, and actor; wrote the screenplay for and directed the motion picture *Un matrimonio interplanetario* (1910).

On this sol in 187 (1960 December 11):

Pathfinders to Mars, Episode 1, “The Imposter” airdate.

On this sol in 210 (2004 March 14):

MER-B Sol 50: *Opportunity* got closer to completing its observations of the rock outcrop. The rover arm, with the alpha particle X-ray spectrometer at the ready, was placed on the rock called Shark's Tooth for a 30-minute observation. The microscopic imager then took a series of pictures of the targets Enamel 1 and Lamination. The focus then switched back to Shark's Tooth for an examination by the Mössbauer spectrometer.

The sol also included many panoramic camera observations of targets with creative names like Patio Rug, Anaconda Snake Den, West Zen Garden and Garter Snake.

On this sol in 210 (2004 March 15):

MER-A Sol 71: *Spirit* began Sol 71 with a morning nap to re-charge after the record-breaking number of activities it accomplished on Sol 70 (210 Sagittarius 10). After that, it was back to work. *Spirit* began by retracting the alpha particle X-ray spectrometer, closing the doors, and imaging the doors with the front hazard avoidance cameras to confirm that they were closed. *Spirit* then proceeded to observe a soil target with the microscopic imager, and it also used the panoramic camera to observe the magnets, do a sky survey and capture a dust devil movie.

Then it was time to drive. *Spirit* completed a 15-meter (49.2 feet) blind drive followed by a 3-meter (9.8 feet) auto-navigation drive around the south rim of Bonneville crater toward a drift named Serpent. Once there, *Spirit* completed post-drive science observations and a miniature thermal emission spectrometer study of the atmosphere, ground and future drive direction.

—NASA/JPL

221 Sagittarius 12 **2024 November 23 03:50 UTC**

Ls 005.3° Curiosity Sol 4372

MSD 53641 Perseverance Sol 1337

On this sol in 174 (1936 July 01):

Part 3 of *The Cometeers* by Jack Williamson appeared in *Astounding Stories* (date approximate).

Interior art by Marco Marchioni.

On this sol in 210 (2004 March 15):

Opportunity reached the first of five targets in its current soil survey. The sol began with a salute to the rover's intended target on the southern face of the crater where it has been exploring since its early sols on Mars.

Before the rover ventured away from the outcrop that has been the focus for the majority of its mission, alpha particle X-ray spectrometer observations were completed on the red rind dubbed Shark's Tooth. The arm was stowed before *Opportunity* scuffed the rock Carousel with its front left wheel. Results of the experiment were imaged as *Opportunity* backed up and prepared for its drive away from the outcrop.

Backing down towards the center of the crater and then arcing around the Challenger Memorial Station, *Opportunity* ultimately drove back up the slope to a position fairly close to the rim. On its way to the current soil target, the rover was also able to image the trench it previously dug on Sol 23 from a different angle.

On this sol in 210 (2004 March 16):

MER-A Sol 72: *Spirit* began the sol taking panoramic camera and miniature thermal emission spectrometer observations of the drift dubbed Serpent before creating the scuff that would reveal the inside material at this location.

Then it was time to get into position. The rover drove about two-and-a-half meters (8.2 feet) to put the left front wheel up onto the drift. It then turned right and left five degrees to dig the left front wheel into the drift. When the shimmy was complete, *Spirit* backed 10 centimeters (3.9 inches) out of the hole. The digging and backing process was repeated four times to thoroughly scuff Serpent's side. Then *Spirit* backed up another meter (3.28 feet) to attain miniature thermal emission spectrometer, panoramic camera and navigation camera observations of the scuffed area. These observations will aid in in-situ target selection. To prepare for the upcoming in-situ work, *Spirit* drove forward 0.4 meters (1.3 feet) for additional imaging, and then drove forward a final 0.45 meters (1.5 feet) to put the scuff in the arm work volume.

Spirit spent the rest of the sol obtaining navigation camera and panoramic camera observations of the intended drive direction around part of the crater rim.

—NASA/JPL

221 Sagittarius 13 **2024 November 24 04:30 UTC**

Ls 005.8° Curiosity Sol 4373

MSD 53642 Perseverance Sol 1338

Perseverance is scheduled to begin its third marsrev of operations at Jezero crater.

On this sol in 154 (1898 November 18):

Andrés Soler was born; Mexican actor; starred in the motion picture *Los platillos voladores (The Flying Saucers)* (1956).

On this sol in 186 (1959 January 26):

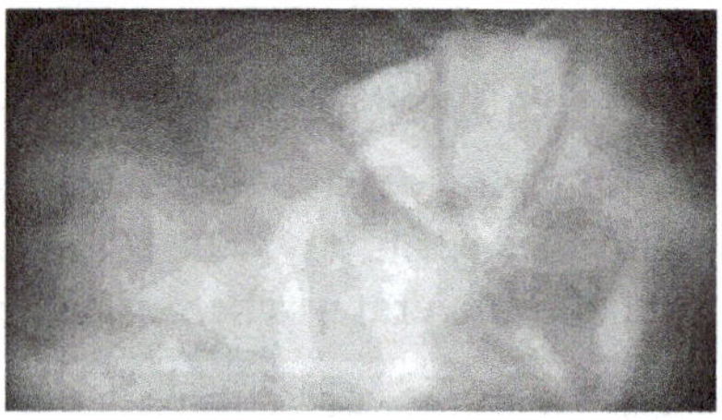

Quatermass and the Pit, Episode 6, "Hob" air date. Chaos reigns in the pit and the long-dormant Martian inheritance begins to assert itself in many people in the area—including Quatermass himself—driving them to a new racial purge. The capsule breaks open and an energy form rises up over London. Roney finds he is immune and strives to defeat it.

—International Movie Database

On this sol in 210 (2004 March 17):

MER-B Sol 52: A full day for *Opportunity,* punctuated by short naps. The rover arm was deployed on the first of five targets in the rover's current soil survey. The soil was examined by all the rover's spectrometers and the panoramic camera.

In the afternoon, the rover moved about 6 meters (about 20 feet) to the second soil target, performing several remote observations along the way. Closing out the sol, *Opportunity's* alpha particle X-ray spectrometer began to inspect the capture magnets on the front of the rover. Clues about Mars' watery history are present even in the smallest dust grains that settle on these magnet arrays.

Images of *Opportunity's* scuff experiment on the rock Carousel reached Earth this sol. Scientists are currently examining the results.

In the coming sols, the rover will travel to the final targets in the soil survey.

On this sol in 210 (2004 March 17):

MER-A Sol 73: *Spirit* spent most of the sol analyzing targets in the drift dubbed "serpent" with the microscopic imager. But before the close-ups began, *Spirit* acquired miniature thermal emission spectrometer inertia measurements on a disturbed area of soil. It also captured panoramic camera images of the scuffed area, dubbed Bear Paw, the wheel that did the digging, and a nearby rock target named White Elephant.

The microscopic imager work began at 11:00 am and was targeted at four points within the scuffed area on Serpent. The targets were given the bear-type names of Polar, Spectacled, Kodiak and Panda. For each target, seven microscopic images were taken to assure proper focus. At the Spectacled and Kodiak targets, a microscopic image with the filter in place was also taken for pseudo-color. Then the Mössbauer was placed on Panda and started an overnight integration. The arm activities were completed by about noon Mars local solar time, and were followed by the second thermal inertia measurement on the disturbed soil.

The rover took a siesta until 1:00 p.m., and then woke up for a series of mini thermal emission spectrometer observations on the rocks named White Elephant, Fruitcake, and Dihedral. Then the third and last thermal inertia measurement was taken on the disturbed soil.

Spirit took another quick nap in the afternoon, and then completed sky measurements with the mini thermal emission spectrometer and panoramic camera.

—NASA/JPL

221 Sagittarius 14 **2024 November 25 05:10 UTC**

Ls 006.3° Curiosity Sol 4374

MSD 53643 Perseverance Sol 1339

Equirria, celebrated on ante dies II Idus Martius on the Roman calendar (analogous to March 14 on the recalibrated Pickering calendar), celebrated with chariot or horse races dedicated to Mars.

On this sol in 140 (1872 July 20):

Walter Clinton Pettee was born; American science fiction cover artist; created the cover art for *A Princess of Mars* (1912) and *The Gods of Mars* (1913) by Edgar Rice Burroughs.

On this sol in 153 (1897 January 01):

Part 4 of *The Martian* by George du Maurier appeared in *Harper's New Monthly Magazine* (date approximate).

Al Martin was born; American television writer and director; wrote seven episodes of the *My Favorite Martian* television series: “Blood Is Thicker Than the Martian,” “Super-Duper Snooper,” “The Sinkable Mrs. Brown,” “Miss Jekyll and Hyde,” “Shake Well and Don’t Use,” “Humbug,” “Mrs. Brown,” and “Won’t You Come Home, Uncle Martin, Won’t You Come Home?”

On this sol in 188 (1962 November 01):

NASA approved two probes for the Mariner Mars 1964 project. The primary goal of the two spacecraft, given the programmatic names Mariner C and Mariner D, was to photograph the Martian surface using a single TV camera fixed on a scan platform that could return up to 22 frames. Ultimately, these would be launched as *Mariner 3* and *Mariner 4*.

On this sol in 210 (2004 March 17):

MER-B Sol 53: *Opportunity* first completed the alpha particle X-ray spectrometer work from yestersol and then continued to study the second soil survey target with the Mössbauer spectrometer.

Following an afternoon nap, the rover used its microscopic imager to get close-up views of the soil. *Opportunity* then made its way to the third soil target, which involved a 5-meter (about 16 feet) drive. The alpha particle X-ray spectrometer was then positioned to examine the magnet arrays. Preparations were also made to conduct a miniature thermal emission spectrometer observation of the Martian sky at dawn on the next sol.

On this sol in 210 (2004 March 18):

MER-A Sol 74: *Spirit* began the morning by completing an alpha particle X-ray spectrometer integration on the target Panda, inside the scuff on Serpent drift. Then *Spirit* placed the alpha particle X-ray spectrometer back down on the target Polar for a 30-minute integration. During that integration, *Spirit* took some images of disturbed soil with the panoramic camera, and acquired some ground temperatures with the miniature thermal emission spectrometer. *Spirit* then switched the tools on its robotic arm to the Mössbauer spectrometer for an hour-long integration on Polar. During that integration, the rover took some sky and ground measurements with the mini thermal emission spectrometer. *Spirit* finished its arm activities for the day by acquiring three microscopic images of Polar and three more of Panda.

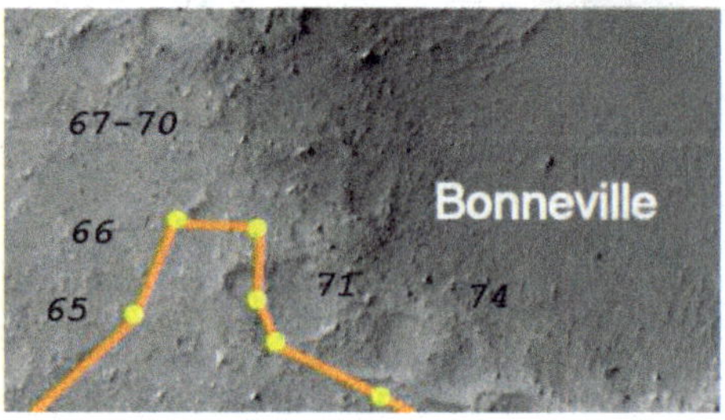

Spirit then took navigation camera and panoramic camera images of the drive directions for planning the Sol 75 (210 Sagittarius 15) traverse. The rover acquired some mini thermal emission spectrometer reconnaissance images and then took a 30-minute siesta before the afternoon *Mars Odyssey* relay pass. During that pass, *Spirit* used the mini thermal emission spectrometer to acquire a sky profile and ground temperature observations.

Starting around 12:35 p.m., *Spirit* made a direct drive of about six meters (19.7 feet) to another section of the Serpent drift complex, called Stub Toe. There the rover repeatedly scuffed the drift and advanced 0.15 meters (half a foot) in a series of five “scuff and drives.”

After the five scuffs and advances were made, *Spirit* roved forward another 3 meters (9.8 feet) and then looked back over its shoulder using the mini thermal emission spectrometer and navigation cameras to analyze the damage. The rover continued along the Bonneville crater rim with a 16-meter direct drive, and then an auto-navigation drive for 9 meters (29.5 feet). *Spirit* completed a final set of drives to set up for a touch and go on Sol 75 at around 2:10 p.m. Mars local solar time. The total amount of driving for Sol 74 was an impressive 34.3 meters (112.5 feet).

—NASA/JPL

221 Sagittarius 15 **2024 November 26 05:49 UTC**

Ls 006.8° Curiosity Sol 4375

MSD 53644 Perseverance Sol 1340

Mamuralia, celebrated on Idus Martius on the Roman calendar (analogous to March 15 on the recalibrated Pickering calendar), a new year festival when a figure called Mamurius Veturius (perhaps the "old Mars" of the old year) was driven out.

On this sol in 210 (2004 March 18):

MER-B Sol 54: *Opportunity* flipped 115 meters (377.3 feet) on its odometer during the latest drives along the current soil survey campaign, surpassing the total drive distance of 1997's *Sojourner* rover. After performing a "touch and go" sequence at the third soil target south of the Challenger Memorial Station, *Opportunity* moved east to its fourth target. There the rover used its wheels to dig a trench that will be studied in coming sols.

The sol started with brief alpha particle X-ray and Mössbauer spectrometer measurements on the soil target known as Coconut 2. These were followed by two sets of microscopic imager shots of Coconut 2 and Chocolate Chip. The rover then stowed its arm and drove.

Remote sensing with the miniature thermal emission spectrometer and the panoramic camera remote sensing was performed before, during, and after the drive and trenching activities. Also, *Opportunity* took additional images with its navigation camera imaging in preparation for next sol's drive to the final site inside the crater.

—NASA/JPL

221 Sagittarius 16 **2024 November 27 06:29 UTC**

Ls 007.3° Curiosity Sol 4376

MSD 53645 Perseverance Sol 1341

On this sol in 158 (1906 June 01):

A Gift from Mars (Part 2 of 3) by William Wallace Cook appeared in The *All-Story Magazine* (date approximate).

221 Sagittarius 17 **2024 November 28 07:08 UTC**

Ls 007.8° Curiosity Sol 4377

MSD 53646 Perseverance Sol 1342

On this sol in 159 (1908 April 19):

"The Great Martian Hoax of 1910 Explained" by Stephen Chalmers appeared in *The New York Times.*

On this sol in 155 (1900 October 10):

Gregory Gaye was born; Russian-American actor; starred in the motion picture *Flying Disc Man from Mars* (1950).

On this sol in 208 (2000 June 17):

A 15-gram basaltic shergottite meteorite designated Dhofar 378 was discovered near Dhofar, Oman.

Roughly three-quarters of all Martian meteorites can be classified as shergottites. They are named after the Shergotty meteorite, which fell at Sherghati, India in 1865. Shergottites are igneous rocks of mafic to ultramafic lithology.

The shergottites appear to have crystallized as recently as 180 million years ago, which is a surprisingly young age considering how ancient the majority of the surface of Mars appears to be, and the small size of Mars itself. Because of this, some have advocated the idea that the shergottites are much older than this. This "Shergottite Age Paradox" remains unsolved and is still an area of active research and debate.

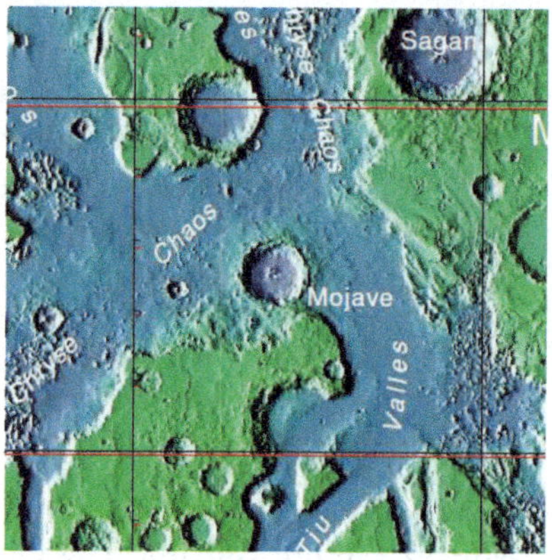

The 3-million-year-old crater Mojave located at 7.5°N 33.1°W, 58.5 km in diameter and the youngest crater of its size on the planet, has been identified as a potential source of these meteorites.

—Wikipedia

On this sol in 210 (2004 March 20):

MER-B Sol 56: NASA's *Opportunity* tried driving uphill out of its landing-site, but slippage prevented success. The rover is healthy, and it later completed a turn to the right and a short drive along the crater's inner slope. Controllers plan to send it on a different route for exiting the crater on Sol 57 (210 Sagittarius 18).

Earlier on Sol 56, *Opportunity* successfully examined a patch of soil dubbed Brian's Choice with its Mössbauer spectrometer, alpha particle X-ray spectrometer and microscopic imager. Following the drive, it made observations with its navigation camera and miniature thermal emission spectrometer.

On this sol in 210 (2004 March 21):

MER-A Sol 77: Before beginning the 0.9-meter (2.95 feet) drive to Mazatzal, *Spirit* analyzed the soil target Soil 1 at its current location with the microscopic imager and Mössbauer spectrometer. During the Mössbauer integration, *Spirit* also took panoramic camera images and performed miniature thermal emission spectrometer analysis of the atmosphere and Mazatzal work area.

At 1:25 p.m. Mars local solar time, *Spirit* completed the Mössbauer integration, took a few microscopic imager images of the impression left on Soil 1 by the Mössbauer spectrometer and then stowed the instrument arm. *Spirit* then proceeded the short distance toward Mazatzal and took hazard avoidance camera images to confirm that its final resting place put the intended rock targets in reach of the instrument arm.

Following the drive, the rover acquired more panoramic camera and mini thermal emission spectrometer observations of the atmosphere, and of interesting areas near the Mazatzal site including targets named Sandbox, Saber, and Darksands.

Spirit finished up Sol 77 by getting the mini thermal emission spectrometer in position for morning observations on Sol 78 (210 Sagittarius 18).

—NASA/JPL

221 Sagittarius 18 **2024 November 29 07:48 UTC**
Ls 008.3° Curiosity Sol 4378
MSD 53647 Perseverance Sol 1343

On this sol in 187 (1960 December 18):

Pathfinders to Mars Episode 2 "Sabotage in Space" airdate.

On this sol in 210 (2004 March 21):

MER-B Sol 57: The image below is the first 360-degree view from the Mars Exploration Rover *Opportunity's* new position outside Eagle Crater, the small crater where the rover landed about two months earlier. Scientists are busy analyzing *Opportunity's* new view of the plains of Meridiani Planum. The plentiful ripples are a clear indication that wind is the primary geologic process currently in effect on the plains. The rover's tracks can be seen leading away from Eagle Crater. At the far left are two depressions—each about a meter (about 3.3 feet) across—that feature bright spots in their centers. One possibility is that the bright material is similar in composition to the rocks in Eagle Crater's outcrop and the surrounding darker material is what's referred to as "lag deposit," or erosional remnants, which are much harder and more difficult to wear away. These twin dimples might be revealing pieces of a larger outcrop that lies beneath. The depression closest to *Opportunity* is whimsically referred to as Homeplate and the one behind it as First Base. The backshell and parachute that helped protect the rover and deliver it safely to the surface of Mars are also visible near the horizon, at the left of the image.

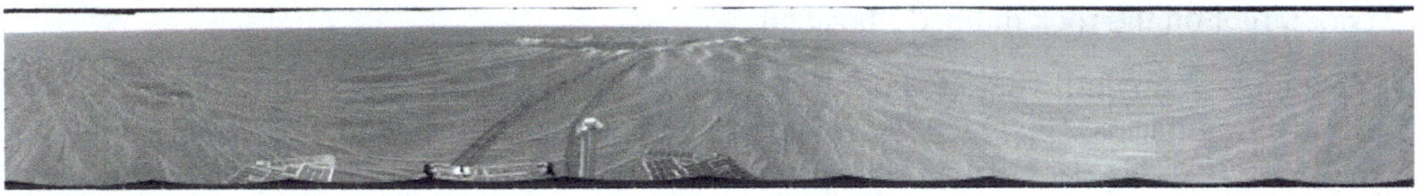

On this sol in 210 (2004 March 22):

MER-A Sol 78: NASA's Mars Exploration Rover *Spirit* woke up at 7:24 a.m. and began a day of observations in preparation for the Sol 79 (210 Sagittarius 19) grind on the rock called Mazatzal.

After waking, *Spirit* warmed-up the mast actuators for some early morning soil and atmosphere miniature thermal emission spectrometer observations. It then went back to sleep before beginning the morning direct-to-earth communication session with the high gain antenna.

At 10:00 a.m. Mars local solar time, *Spirit* began analyzing the soil targets Saber and Sandbox with the mini thermal emission spectrometer. It also took panoramic camera images of Skull and Saber. Then it was time to unfold the instrument arm to capture microscopic imager images of three targets on Mazatzal: Arizona, Illinois, and New York. The New York target was further analyzed with a 17-hour Mössbauer spectrometer integration.

While the Mössbauer was integrating, *Spirit* proceeded to execute several mini thermal emission spectrometer and panoramic camera observations of interesting features in the surrounding area. The observations included images of Bonneville crater, Saber, Sandbox, and Orange Beach.

Spirit had completed all these activities by 2:40 p.m. Mars local solar time and then took a siesta until the afternoon *Mars Odyssey* UHF pass. During that pass, the rover captured mini thermal emission spectrometer ground and sky images. Before shutting down at 5 p.m. Mars local solar time, *Spirit* positioned the panoramic camera for a nighttime observation of the moon Deimos.

—NASA/JPL

221 Sagittarius 19 **2024 November 30 08:27 UTC**
Ls 008.8° Curiosity Sol 4379
MSD 53648 Perseverance Sol 1344

On this sol in 156 (1902 August 31):

Reginald Dandy (John Garrick) was born; British-American actor; starred in the motion picture *Just Imagine* (1930).

On this sol in 210 (2004 March 22):

MER-B Sol 58: *Opportunity* began to image a panoramic mosaic of the plains on this sol.

On this sol in 210 (2004 March 23):

MER-A Sol 79: NASA's Mars Exploration Rover *Spirit* awakened at 9:35 a.m. An early morning review of the alpha particle X-ray spectrometer data revealed that the instrument's doors were not fully open and that the tool did not completely engage at the intended New York target on the rock dubbed Mazatzal.

Spirit took a nap until 12:45 to conserve power for the upcoming grind on Mazatzal on Sol 81 (210 Sagittarius 21). Once the rover woke up, it began the sequences of brushing and analyzing two targets, New York and Illinois, on Mazatzal. Each target was brushed with the rock abrasion tool and then imaged with the microscopic imager and panoramic camera. The entire sequence ended with a Mössbauer spectrometer integration on the New York target.

—NASA/JPL

221 Sagittarius 20 **2024 December 01 09:07 UTC**
Ls 009.3° Curiosity Sol 4380
MSD 53649 Perseverance Sol 1345

On this sol in 153 (1897 January 08):

Dennis Wheatley was born; British writer of thrillers and occult novels; author of *Star of Ill-Omen* (1952).

On this sol in 210 (2004 March 23):

MER-B Sol 59: Opportunity spent the sol placing the Mössbauer spectrometer on the bright material it approached yestersol, and conducting more remote sensing observations.

This relatively light workload allowed the rover to recover energy for the next sol's activities. Those will include completing an alpha particle X-ray spectrometer read on the same soil target and initiating the panoramic mosaic image from the rover's current position.

On this sol in 210 (2004 March 24):

MER-A Sol 80: NASA's Mars Exploration Rover *Spirit* repeated overnight measurements of Illinois and New York, two targets on the rock Mazatzal. The measurements needed to be repeated because the alpha particle X-ray spectrometer's doors inadvertently did not open during the prior sol.

Mazatzal is one of an apparent class of light-toned rocks, which may be common in the area where *Spirit* landed in Gusev. This rock appears to be a ventifact, which means it may have been carved by the steady winds that scientists know come from the northwest into the top area of this crater rim.

—NASA/JPL

221 Sagittarius 21 **2024 December 02 09:47 UTC**
Ls 009.8° Curiosity Sol 4381
MSD 53650 Perseverance Sol 1346

On this sol in 210 (2004 March 24):

MER-B Sol 60: Opportunity obtained this panorama, dubbed Lion King, on Sols 58 and 60 (210 Sagittarius 19 and 21) of the mission as the rover was perched at the lip of Eagle Crater, majestically looking down into its former home. It is the largest panorama yet obtained by either rover.

The full panoramic image was taken in eight segments using six filters per segment, for a total of 558 images and more than 75 megabytes of data. This enhanced color composite was assembled from the infrared (750 nanometer), green (530 nanometer), and violet (430 nanometer) filters. Additional lower elevation tiers were added relative to other panoramas to ensure that the entire crater was covered in the mosaic.

On this sol in 210 (2004 March 35):

MER-A Sol 81: *Spirit* began the sol by stopping the alpha particle X-ray spectrometer integration and then imaging the instrument's placement with the hazard avoidance camera. The rest of this sol was all about grinding into the target New York on the rock named Mazatzal.

The rock abrasion tool operated on the New York target for three hours and forty-five minutes and created an impression in the rock that was 3.79 millimeters (.15 inches) deep. The angular shape of Mazatzal and the fact that the rock is a little harder than previously abraded rocks allowed the flatter side of the circular target to receive a more intense grind.

Spirit's odometer now reads: 492 meters - more than a quarter of a mile.

—NASA/JPL

221 Sagittarius 22 **2024 December 03 10:26 UTC**

Ls 010.3° Curiosity Sol 4382

MSD 53651 Perseverance Sol 1347

Earth departure opportunity to Mars via a Type II trajectory.

Earth departure opportunity to Mars via a Type IV trajectory.

On this sol in 153 (1897 January 09):

Oscar Jerome Friend was born; American literary agent, editor, and writer; author of "The Worms Turn" (1940).

On this sol in 210 (2004 March 26):

MER-A Sol 82: *Spirit* spent most of the sol analyzing the rock abrasion tool impression with the microscopic imager, Mössbauer spectrometer and alpha particle X-ray spectrometer. The rover also placed the microscopic imager over a clean surface to the upper left of the ratted area and take some images.

—NASA/JPL

221 Sagittarius 23 **2024 December 04 11:06 UTC**

Ls 010.8° Curiosity Sol 4383
MSD 53652 Perseverance Sol 1348

On this sol in 154 (1898 November 29):

Clive Staples Lewis was born; British writer and lay theologian; author of *Out of the Silent Planet* (1938).

221 Sagittarius 24 **2024 December 05 11:45 UTC**

Ls 011.3° Curiosity Sol 4384
MSD 53653 Perseverance Sol 1349

On this sol in 210 (2004 March 28):

MER-A Sol 84: *Spirit* surpassed *Sojourner's* endurance record for rover operation on the surface of Mars.

The angular nature of the rock called Mazatzal required some extra rodent power over the weekend. The latest grind by *Spirit's* rock abrasion tool (the RAT) resulted in the deepest human-made hole on another planet—nearly 8 millimeters (0.31 inches.

Spirit's 84th sol was planned as a sol of investigation. The miniature thermal emission spectrometer and panoramic camera made successful observations of the crater informally named Bonneville, but planned operations of the rover arm were not executed due to a switch on the Mössbauer spectrometer getting momentarily stuck. After a successful Mössbauer integration, the instrument was pulled back from Mazatzal, but one of two contact switches did not indicate a no-contact condition. Although the instrument had been retracted, the rover's software interpreted this as an unexpected collision of the spectrometer with an object, so it terminated any further arm operations. The stuck switch flipped about three minutes later but the rover is programmed to wait until the false error is cleared by mission control.

—NASA/JPL

221 Sagittarius 25 **2024 December 06 12:25 UTC**

Ls 011.8° Curiosity Sol 4385

MSD 53654 Perseverance Sol 1350

On this sol in 175 (1938 June 01):

"A Summons from Mars" by John Russell Fearn appeared in *Amazing Stories* (date approximate). In a weird message from Mars, Eric Sanders is told that he has been betrothed to Yana. Torn between his duty to science and tradition, and his love for Sonia Benson, Sanders makes a momentous decision.

Interior art by Harold Welch.

"Escape Through Space" by Ross Rocklynne also appeared in Amazing Stories (date approximate). The Martian dynasty overthrown, the ex-princess is exiled to Earth under escort by Larry Sharon. Pursued by treacherous revolutionists, the unhappy pair face the deadly "Boiling Zone" of the sun.

Interior art by Jay Jackson.

On this sol in 187 (1960 December 25):

Pathfinders to Mars Episode 3 "The Hostage" airdate.

On this sol in 210 (2004 March 29):

MER-B Sol 64: The *Opportunity* rover team analyzed the results engineering activities run to investigate an error message they received from the rover on Sol 63 (210 Sagittarius 24).

A problem with a secondary memory file was isolated and resolved. Just as an ordinary computer disk can have corrupted sections, a corrupted file in an area where rover commands are addressed and stored has been identified. Engineers have identified the location of the problem within the memory and figuratively fenced it off, containing it and preventing it from harming any future command sequences. This minor issue has not impeded the rover from resuming normal science operations on the next sol.

The rover is currently at the rock dubbed Bounce. *Opportunity* met this rock once before; while still cloaked in its protective lander and airbags, the rover bounced on the rock while on its way to a safe landing in Eagle Crater. Miniature thermal emission spectrometer observations have shown Bounce is rich in hematite. In the coming sols, the rover's other spectrometers will examine the rock before the rock abrasion tool grinds into a designated target.

MER-A Sol 85: As planned, *Spirit* began the sol by receiving the uplink and then taking a one-hour nap. After waking, the rover took panoramic camera images of the rock abrasion tool and of the ratted hole in Mazatzal. Before the panoramic camera work was done, *Spirit* took some final shots of Bonneville crater. Some of those images might contribute to a super-resolution image of the heatshield remnants on the far wall. *Spirit* also took some images to try to catch a dust devil in action.

A little after 2 p.m., the last miniature thermal emission spectrometer sections of Bonneville crater were acquired, followed by a set of panoramic camera images of the drive direction. In the late afternoon, *Spirit* used the mini thermal emission spectrometer to acquire ground and sky stares.

After the panoramic camera activity, *Spirit* used the microscopic imager to take a 5-position pseudo-color mosaic of the Mazatzal rock abrasion tool hole. Then the Mössbauer spectrometer was placed in the hole and began an overnight integration.

Since the rock abrasion tool completed a full-circle grind into the New York and Brooklyn targets on the rock Mazatzal, it was time for Spirit to do some analysis. *Spirit* spent much of Sol 85 successfully operating the instruments on its robotic arm to take a more detailed look inside Mazatzal.

Although all the operations were completed successfully on Mars, the rover team spent most of the morning and afternoon on Earth worrying. After the team sent the uplink to *Spirit,* they waited for the standard "beep" that confirms the sequence reached *Spirit* and was activated. This beep, and an expected one 10 minutes later were not acquired, and engineers proceeded to trouble-shoot what might have gone wrong. No errors could be found, and finally a successful afternoon *Mars Odyssey* communications pass provided 75 megabits of data, indicating that all the sequences were in fact onboard the rover and that all the planned sol activities had completed successfully. Like worried parents, the rover team members breathed a collective sigh of relief, and are now looking into possible causes of the failure to detect the beep.

—NASA/JPL

221 Sagittarius 26 **2024 December 07 13:05 UTC**
Ls 012.3° Curiosity Sol 4386
MSD 53655 Perseverance Sol 1351

On this sol in 210 (2004 March 30):

MER-B Sol 65: During the Martian morning, *Opportunity's* miniature thermal emission spectrometer and panoramic camera studied the atmosphere. Bounce rock was imaged by the panoramic camera.

Opportunity's instrument arm was then deployed to get a close-up view of Bounce using the microscopic imager. The rock abrasion tool team used some of these images to identify the exact target for next sol's grinding operation. The Mössbauer spectrometer was then placed on a designated target on the rock for an overnight integration.

In the afternoon, *Opportunity* took navigation and panoramic camera images and completed more miniature thermal emission spectrometer science.

MER-B Sol 86: *Spirit* began the sol by waking up and heating the panoramic mast assembly to complete sky and ground stares with the miniature thermal emission spectrometer. *Spirit* completed the alpha particle X-ray spectrometer integration on the hole made by the rock abrasion tool and then took a 45-minute nap.

Once *Spirit* woke up, it began the 6-position rock abrasion tool brush mosaic on the target Missouri on the rock called Mazatzal. Once this was completed successfully, the rover's arm was stowed.

Spirit then rolled backwards 0.9 meters (2.95 feet) to correctly position itself to acquire mini thermal emission spectrometer imaging of the newly brushed mosaic, and the previously ground hole. In addition, *Spirit* took sky and ground stares and panoramic camera images of the upcoming drive direction. The sol ended with mini thermal emission spectrometer stares at the Columbia Hills and an afternoon pass by NASA's *Mars Odyssey* orbiter.

—NASA/JPL

221 Sagittarius 27 **2024 December 08 13:44 UTC**

Ls 012.8° Curiosity Sol 4387

MSD 53656 Perseverance Sol 1352

Predicted meteor shower associated with parent body C/1979 Y1 Bradfield: Martian right ascension 186.3°, declination -30.2°. Impacting the southern hemisphere, morning side.

On this sol in 184 (1955 May 08):

The *X Minus One* presentation of the radio drama "Mars Is Heaven" by Ray Bradbury was broadcast on KNBC in San Francisco. The first Earth visitors to Mars discover a replica of small town America, just as they remembered it from their childhoods.

—International Movie Database

On this sol in 210 (2004 March 31):

MER-B Sol 66: *Opportunity's* rock abrasion tool ground into Bounce for just over two hours, producing a 6.44-millimeter (0.25 inch) hole that will allow the rover's spectrometers to analyze the rock's chemical composition.

The Martian morning began with the alpha particle X-ray spectrometer observing a target called Glanz2 on Bounce. Miniature thermal emission spectrometer measurements of the ground and sky followed.

The rock abrasion tool was then placed on the target dubbed "Case." After the grind, the Mössbauer spectrometer was placed on the hole for an overnight integration.

In the afternoon, the rover also had time to complete more atmospheric science with its panoramic camera and miniature thermal emission spectrometer.

MER-A Sol 87: *Spirit* began the sol with some morning atmospheric science, and then took a last look at the rock Mazatzal with the panoramic and navigation cameras. Then the rover was off, traveling 36.5 meters (119.8 feet) down the side of Bonneville Crater headed south toward the Columbia Hills.

Following the drive, *Spirit* took navigation and panoramic camera pictures in her drive direction and performed atmospheric and soil science with the panoramic camera and mini thermal emission spectrometer.

The drive was a combination of "blind" and autonomous navigation roving. The blind segments of the drive are used when rover planners can see all possible hazards and command the rover to just "go." The autonomous navigation portion allows the rover to make decisions based on the terrain presented. While the blind segments of the Sol 87 drive were successful, the second to the last autonomous navigation sequence did not complete in the allotted time, causing a drive "goal" error. As a result, *Spirit* was not able to execute the complete commanded drive, and roved 36.5 meters (119.8 feet) of the 65-meter (213.3 feet) planned drive.

—NASA/JPL

221 Sagittarius 28 **2024 December 09 14:24 UTC**
Ls 013.3° Curiosity Sol 4388
MSD 53657 Perseverance Sol 1353

Predicted meteor shower associated with parent body C/1942 X1 Whipple–Fedtke–Tevzadze: Martian right ascension 15.8°, declination -24.9°.

On this sol in 210 (2004 April 01):

MER-B Sol 67: *Opportunity* continued to examine Bounce with the microscopic imager and the Mössbauer and alpha particle X-ray spectrometers.

During the Martian morning, the Mössbauer spectrometer was turned off before atmospheric science was conducted with the miniature thermal emission spectrometer and the panoramic camera.

The afternoon hours were dedicated to intensive study of a handful of targets on Bounce, including the impression ground by the rock abrasion tool on Sol 66 (210 Sagittarius 27).

MER-A Sol 88: *Spirit* spent most of the sol driving toward the Columbia Hills. Before beginning the drive, *Spirit* acquired some pre-drive remote sensing, which included panoramic camera photometry and a mini thermal emission spectrometer stare of the rock called Carlsbad.

Spirit then began the 35-meter (114.8 feet) combination directed and autonomous navigation drive down the rocky, ejecta-covered side of Bonneville Crater. Fifteen meters (49.2 feet) of the drive were directed by rover planners and did not require the rover to use its hazard avoidance software. The remaining 20 meters (65.6 feet) were navigated by *Spirit* autonomously and did cause the rover to make some back and forth adjustments as it avoided what it perceived to be a depression hazard in its path.

—NASA/JPL

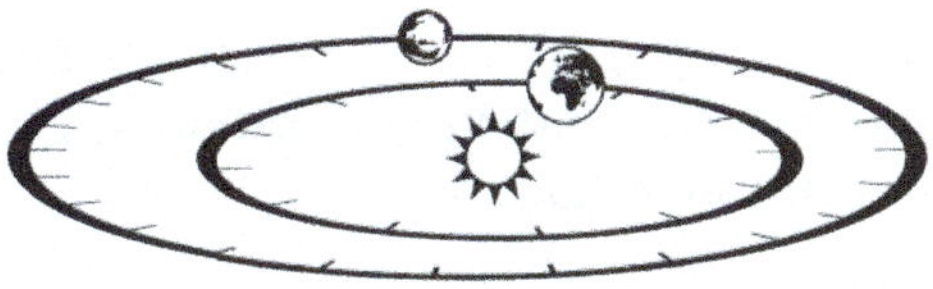

221 DHANUS

As seen from Mars, the Sun traverses the eastern sector of the constellation Sagittarius, entering Capricornus at the end of the month. The name Dhanus (धनुष), the Sanskrit word for “bow,” is taken from a month in the Hindu solar calendar that originally occurred when the Sun passed through the constellation Sagittarius as seen from Earth. Shiva Dhanush (शिव धनुष) is the bow of Lord Shiva.

The median one-way communications delay during this month is 5 minutes 39 seconds.

221 Dhanus 01 **2024 December 10 15:03 UTC**

Ls 013.8° Curiosity Sol 4389

MSD 53658 Perseverance Sol 1354

Approximately the first sol of the unnamed month between September and October on the Heinlein calendar.

Predicted meteor shower associated with parent body C/1942 X1 Whipple–Fedtke–Tevzadze: Martian right ascension 15.8°, declination -24.9°.

MER-B Sol 68: *Opportunity* backed away from Bounce, then re-approached the rock in preparation for an alpha particle X-ray spectrometer read on the right side of it. *Opportunity* made observations with its miniature thermal emission spectrometer during the Martian morning. Then it took set of microscope images before backing away from the rock. More images were taken from that vantage point before *Opportunity* made a 10-degree turn in place and drove the 0.85 meters (2.8 feet) back to Bounce.

—NASA/JPL

221 Dhanus 02 **2024 December 11 15:43 UTC**

Ls 014.3° Curiosity Sol 4390

MSD 53659 Perseverance Sol 1355

On this sol in 173 (1934 September 01):

“Dragon’s Teeth” by Wallace West appeared in *Astounding Stories* (date approximate). In the past humanity developed a high civilization on the continent of Mu. which sank into the sea. Before the submergence, however, some natives of Mu fled to Mars, where they settled; other Muans, surviving in the colonies, ultimately founded the kingdoms of Atlantis and Egypt. Both groups lost much of the ancient science, but are still far ahead of us. Now, five thousand years later, a Martian expedition led by King Poseidon has landed in Atlantis, intent on conquering it.

—Bleiler and Bleiler, 1998

Interior art by Elliot Dold.

“Famine on Mars” by Frank K. Kelly also appeared in *Astounding Stories* (date approximate). Portrays a dying Martian civilization in its struggle to survive. A tale about altruism versus greed.

Interior art by Marco Marchioni.

On this sol in 210 (2004 April 03):

MER-B Sol 69: *Opportunity* completed its instrument arm work on Bounce. It also examined soil targets with its microscopic imager and Mössbauer spectrometer.

—NASA/JPL

221 Dhanus 03 **2024 December 12 16:23 UTC**

Ls 014.8° Curiosity Sol 4391
MSD 53660 Perseverance Sol 1356

On this sol in 210 (2004 April 04):

MER-B Sol 70: *Opportunity* dashed away from the rim of its Eagle Crater landing site. The roughly 100-meter (about 328 feet) drive led the rover to a target area dubbed Anatolia, along a sinuous crack in the plains of Meridiani Planum defined by deep impressions in the sand sprinkled with Eagle Crater-like rocks. In the coming sols *Opportunity* will further investigate the rocks in this mini-outcrop.

Before leaving the vicinity of Eagle Crater, *Opportunity* performed a maneuver on Bounce rock lightheartedly called “crush and go” by the rover engineers. In order to gather further information about the rock’s hardness, the intentional drive over Bounce was an attempt to fracture it. The science team is awaiting images from the rover’s rear hazard avoidance camera to see the results.

MER-A Sol 91: *Spirit’s* primary mission concluded.

Finishing 90 sols of surface operations since landing day marked completion of the last of the official success criteria for *Spirit’s* prime mission. The rover team at JPL had checked off the next-to-last box for mission success two days earlier, when a drive of 50.2 meters (165 feet) took *Spirit’s* total travel distance over the 600-meter (1,969 feet) mark.

Sol 91 started with some remote sensing observations of the sky and ground as well as navigation camera images of the landscape to the east. Then the rover completed miniature thermal emission spectrometer ground surveys and imaged the sky and ground with the panoramic camera. After a short nap, *Spirit* acquired some pre-drive imaging including a super-spectral look at an interesting spot in front of the rover.

Early in the Martian afternoon, *Spirit* began a 1.35-meter (4.4-foot) drive to get closer to a rock called Route 66. Once the drive was finished, the rover analyzed the instrument deployment device’s work volume with hazard-avoidance camera images and a stare by the miniature thermal emission spectrometer. A quick adjustment of 0.8 meters (2.6 feet) put the rover in perfect position and completed the drive.

Spirit spent the afternoon taking a systematic soil survey with the panoramic camera, a 13-filter image of the destination informally named Columbia Hills, and acquiring miniature thermal emission spectrometer data of the same locations.

—NASA/JPL

221 Dhanus 04 **2024 December 13 17:02 UTC**
Ls 015.2° Curiosity Sol 4392
MSD 53661 Perseverance Sol 1357

On this sol in 187 (1961 January 01):

Pathfinders to Mars Episode 4 "Lichens!" airdate.

On this sol in 210 (2004 April 05):

MER-A Sol 92: *Spirit's* extended mission commenced. The rover awoke and completed some early morning panoramic camera sky and ground measurements. *Spirit* also took a look at the capture and filter magnets with the panoramic camera prior to taking a short mid-morning nap. Upon wake-up around 12:30 p.m., the rover opened the doors on the alpha particle X-ray spectrometer and took 3 images of each magnet. *Spirit* also placed the Mössbauer spectrometer on the capture magnet and began an integration.

In the afternoon, *Spirit* completed coordinated observations with the thermal emission spectrometer instrument on NASA's *Mars Global Surveyor* orbiter. The observations involved miniature thermal emission spectrometer pre-flight, simultaneous, and post-flight sky and ground measurements. *Spirit* also collected a panoramic camera opacity observation.

—NASA/JPL

221 Dhanus 05 **2024 December 14 17:42 UTC**
Ls 015.7° Curiosity Sol 4393
MSD 53662 Perseverance Sol 1358

On this sol in 188 (1962 November 21):

NASA approved Voyager, a pair of unmanned orbiter/lander missions to be launched on Saturn IB-Centaur launch vehicles to Venus in 1967 and to Mars in 1969.

On this sol in 210 (2004 April 06):

MER-B Sol 72: The rover drove around the sinuous trough in a long dogleg pattern. Remote sensing to examine the crevice was conducted on the 50-meter (164 feet) drive to its ultimate position for the sol, at the northeast extreme of Anatolia.

MER-A Sol 93: *Spirit* began the sol by heating the high gain antenna. *Spirit* then took some calibration measurements and navigation camera imaging of the alpha particle X-ray spectrometer's placement on the magnet.

Around 10:30 a.m., *Spirit* moved its instrument deployment device out of the way to take hazard avoidance camera images of the rock target called Route 66. The rover then operated the Mössbauer spectrometer for a four-hour integration. During the integration, *Spirit* captured its current location with a panoramic camera mosaic toward the Columbia Hills and the rover tracks.

After a short nap, *Spirit* took mini thermal emission spectrometer observations of the rover tracks, and rock targets called Everest and Pisa. the rover completed the Mössbauer integration and then placed it on Route 66 with the alpha particle X-ray spectrometer pointed up to the sky. The instrument deployment device will remain in this position during the flight software load time period, allowing for long integrations of the Mössbauer on Route 66 and the alpha particle X-ray spectrometer on the sky.

—NASA/JPL

221 Dhanus 06 **2024 December 15 18:21 UTC**
Ls 016.2° Curiosity Sol 4394
MSD 53663 Perseverance Sol 1359

On this sol in 159 (1908 May 07):

Nature published a letter by E.H. Hankin in which he suggested that the linear features on Mars were the extent of a primitive life form which had spread across the planet.

On this sol in 210 (2004 April 07):

MER-B Sol 73: *Opportunity* was on the move again. The rover toured and examined the trough remotely.

MER-A Sol 94: *Spirit* began a four-sol stand-down. During this time, the rover will receive a flight software update that should make its remaining Martian days even safer and more productive. The upload will run through Sol 97 (210 Dhanus 09) with a rover re-boot on Sol 98 (210 Dhanus 10). The rover will be receiving the same update package in upcoming sols.

Spirit is currently parked in front of the rock called Route 66, and will remain there for the duration of the flight software update, with the Mössbauer spectrometer integrating on the rock, and the alpha particle X-ray spectrometer pointed up to the sky.

The flight software update package includes three key changes. First is an update to the autonomous navigation software that will allow *Spirit* to travel longer distances autonomously over the extremely rocky Gusev Crater terrain. The current autonomous navigation software sometimes gets stuck when it detects a hazard that it can't navigate around. The new version will allow *Spirit* to turn in place to find the best possible path.

The second part of the flight software update will allow *Spirit* to recover more easily from an anomaly like the one that occurred on Sol 18. Although operational processes and software have already been updated to prevent something like this from ever happening again, engineers have included additional safety nets in the software that would allow the rover to autonomously react to a similar anomaly and recover to a more stable state.

The third portion of the update is specific to *Opportunity* and is intended to mitigate against energy loss associated with the stuck heater on *Opportunity's* instrument deployment device. The fix allows rover planners to put the rover in a deep sleep mode, where the batteries are totally removed from being able to power the stuck switch. Therefore, with no power reaching the stuck heater switch, the *Opportunity* rover battery will not be drained.

Rover controllers will not initiate the deep sleep capability on *Spirit* unless it becomes necessary.

—NASA/JPL

221 Dhanus 07 **2024 December 16 19:01 UTC**

Ls 016.7° Curiosity Sol 4395

MSD 53664 Perseverance Sol 1360

On this sol in 210 (2004 April 08):

MER-A Sol 95: *Spirit* remained stood down while receiving a flight software update package.

—NASA/JPL

221 Dhanus 08 **2024 December 17 19:40 UTC**

Ls 017.2° Curiosity Sol 4396

MSD 53665 Perseverance Sol 1361

On this sol in 210 (2004 April 09):

MER-B Sol 75: *Opportunity* began a four-sol stand-down. During this time, the rover will receive a flight software update that should make its remaining sols even safer and more productive. The upload will run through Sol 78 (210 Dhanus 11) with a rover re-boot on Sol 79 (210 Dhanus 12).

Opportunity is currently parked near the trench it dug on its 73rd sol. It will remain there for the duration of the flight software update. To keep the battery charge high, engineers are not planning to integrate the rover spectrometers on a target in the trailings of the trench during the flight software update.

The flight software update package includes three key changes. First is an update to the autonomous navigation software that will allow both rovers to travel longer distances autonomously. The current autonomous navigation software sometimes gets stuck when it detects a hazard that it can't navigate around. The new version will allow the rovers to turn in place to find the best possible path.

The second part of the flight software update will allow the rovers to recover more easily from an anomaly like the one that occurred on *Spirit's* Sol 18 (209 Scorpius 14). Although operational processes and software have already been updated to prevent something like this from ever happening again, engineers have included additional safety nets in the software that would allow the rovers to autonomously react to a similar anomaly and recover to a more stable state.

The third portion of the update is specific to *Opportunity* and is intended to mitigate against energy loss associated with the stuck heater on *Opportunity's* instrument deployment device. The fix allows rover planners to put the rover in a deep sleep mode, where the batteries are totally removed from being able to power the stuck switch. Therefore, with no power reaching the stuck heater switch, the *Opportunity* rover battery will not be drained. Rover controllers will not initiate the deep sleep capability on *Spirit* unless it becomes necessary.

MER-A Sol 96: *Spirit* remained stood down while receiving a flight software update package.

—NASA/JPL

221 Dhanus 09 **2024 December 18 20:20 UTC**
Ls 017.7° Curiosity Sol 4397
MSD 53666 Perseverance Sol 1362

On this sol in 185 (1957 April 03):

The *X Minus One* presentation of the radio drama "Martian Sam" by Ernest Kinoy was broadcast on WRCA in New York City. A baseball coach discovers a rookie ballplayer of colossal ability. The tyro can hit, field, run and throw as well as anyone in the history of the game. The only problem is...he's a Martian!

On this sol in 210 (2004 April 10):

MER-B Sol 76: *Opportunity* remained stood down while receiving a flight software update package.

MER-A Sol 97: *Spirit* remained stood down while receiving a flight software update package.

—NASA/JPL

221 Dhanus 10 **2024 December 19 21:00 UTC**
Ls 018.2° Curiosity Sol 4398
MSD 53667 Perseverance Sol 1363

On this sol in 187 (1961 January 08):

Pathfinders to Mars Episode 5 "Zero Hour on the Red Planet" airdate.

On this sol in 210 (2004 April 11):

MER-B Sol 77: *Opportunity* remained stood down while receiving a flight software update package.

MER-A Sol 98: The good news is that *Spirit's* "mind" is updated and operating as expected.

After confirmation that the new software was running as expected, engineers ran some clean-up activities and breathed a sigh of satisfaction.

Controllers gave the go to reboot the rover's computer, which would then run the new software during the morning of Sol 98. The command was sent, and a little over a half hour later, engineers saw the carrier beep that indicated that the command was received. *Spirit* went to sleep for several minutes after that, and then woke up, rebooting into the new software. It then initiated a high-gain antenna session at 12:30 p.m. local solar time, and engineers saw that the new version of the rover software was running properly.

The new software provides several improvements. It enhances the rovers' mobility, and should allow *Spirit* to go much farther each sol by reducing how often it has to take images and generate new three-dimensional maps. A deep-sleep mode has been added, mainly to resolve the heater that is stuck in the "on" position on the rover. However, someday *Spirit* may find a use for this mode as well. The new software also mitigates the memory problem *Spirit* had on Sol 18 (209 Scorpius 14), and seems to be making a difference already. *Spirit's* unallocated memory jumped from 2.0 megabytes to 3.3 megabytes after the new software compacted the flash directories when it booted.

—NASA/JPL

221 Dhanus 11 **2024 December 20 21:39 UTC**
Ls 018.7° Curiosity Sol 4399
MSD 53668 Perseverance Sol 1364

On this sol in 210 (2004 April 12):

MER-B Sol 78: Opportunity remained largely stood down while receiving a flight software update package. The rover took daytime readings with its Mössbauer spectrometer on Jeff's Choice—a soil target in the tailings of the trench that the vehicle dug on Sol 73 (210 Dhanus 06).

MER-A Sol 99: *Spirit* began the sol by doing a systematic ground survey with the miniature thermal emission spectrometer. After that, the rover completed a six-position brush mosaic on the rock Route 66 with the rock abrasion tool. Once the brushing was complete, *Spirit* analyzed the area with the microscopic imager and Mössbauer spectrometer.

The afternoon science for the sol included imaging of rocks called Back Lot and Cameo with the panoramic camera and the miniature thermal emission spectrometer. *Spirit* took a short nap and woke up for the afternoon *Mars Odyssey* pass.

During the Martian night, *Spirit* changed from using the Mössbauer spectrometer to using the alpha particle X-ray spectrometer.

—NASA/JPL

221 Dhanus 12 **2024 December 21 22:19 UTC**
Ls 019.2° Curiosity Sol 4400
MSD 53669 Perseverance Sol 1365

On this sol in 210 (2004 April 13):

MER-B Sol 79: *Opportunity* performed a free-air integration of its alpha particle X-ray spectrometer. This procedure measures the effect of the Mössbauer's radiation source on the alpha particle X-ray spectrometer's sensor, allowing the science team to subtract out the Mössbauer influence for an accurate calibration.

MER-A Sol 100: *Spirit* began the sol by closing the doors on the alpha particle X-ray spectrometer, stowing the robotic arm and backing up from the rock called Route 66 in preparation for an afternoon drive. Before taking off, *Spirit* used the miniature thermal emission spectrometer to examine a patch of Route 66's surface that had been scrubbed in a daisy-shaped mosaic of brushings by the rock abrasion tool.

This composite image from the panoramic camera on *Spirit* gives an approximately true-color rendering of a daisy-like pattern of brushed circles that *Spirit* produced on a rock called Route 66. *Spirit* used the rock abrasion tool to complete this 6-position "RAT daisy" on Sol 99 (210 Dhanus 11). It took this image on Sol 100.

The purpose for these large brushings is to create a large enough patch of treated surface area for the miniature thermal emission spectrometer to analyze. Scientists had previously conducted a brushing like this one on the rock "Mazatzal." The brushed area of Route 66 looks very different from the brushed area of Mazatzal, leading scientists to think that the rocks—although both light in tone—actually have different coating types.

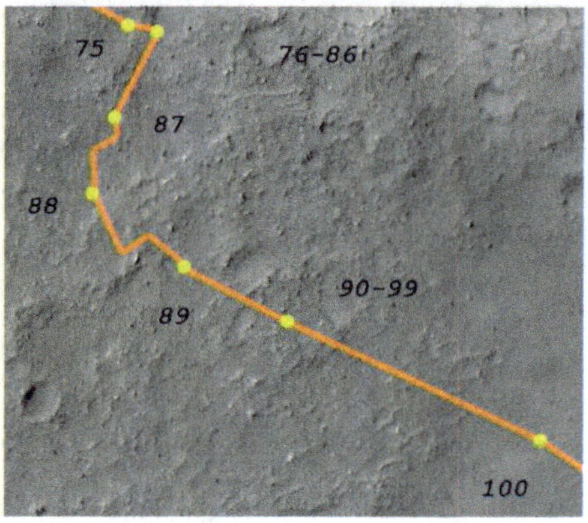

In the early Martian afternoon, the rover began a 64-meter (210 feet) drive, *Spirit's* longest one-sol drive so far. The final 24 meters (78.7 feet) of the drive were navigated with the enhanced autonomous navigation capabilities of *Spirit's* newly uploaded software. The new software has nearly doubled the meters-per-hour rate to 32 (105 feet-per-hour).

After the traverse, *Spirit* completed post-drive imaging and used the miniature thermal emission spectrometer on sky and ground targets.

With 100 sols under her belt and 706.5 meters (0.44 miles) on her odometer, *Spirit* keeps on roving toward the Columbia Hills.

—NASA/JPL

221 Dhanus 13 **2024 December 22 22:58 UTC**

Ls 019.6° Curiosity Sol 4401
MSD 53670 Perseverance Sol 1366

On this sol in 210 (2004 April 14):

MER-B Sol 80: *Opportunity* spent the sol examining the trench it dug on Sol 73 (210 Dhanus 06). The rover's microscopic imager got close-up views of the targets called Jeff's Choice, RipX, Jack Russell, Beagle Burrow and NewRipX in the trench. The navigation and panoramic cameras shot images in *Opportunity's* drive direction toward Endurance Crater (right).

The rover's spectrometers gathered data at several of the soil targets. Atmospheric data was collected by the miniature thermal emission spectrometer and panoramic camera.

MER-A Sol 101: Right Navigation Camera non-linearized full frame experiment data record (EDR, right) acquired of *Spirit's* mission to Gusev Crater at approximately 13:10:25.

—NASA/JPL

221 Dhanus 14 **2024 December 23 23:38 UTC**

Ls 020.1° Curiosity Sol 4402
MSD 53671 Perseverance Sol 1367

Agonalia (or Agonium Martiale), celebrated on ante dies XVI Kalendas Aprilis on the Roman calendar (analogous to March 42 on the revised Pickering calendar), an obscure type of observance for Mars.

On this sol in 174 (1936 August 01):

"Subconscious" by John Russell Fearn appeared in *Amazing Stories* (date approximate). We are to be taken millions of miles from the earth we know so well, are to get the story of the transactions on a distant planet, with a reference to some mysterious celestial phenomena and a stroke of lightning….

Interior art by Leo Morey.

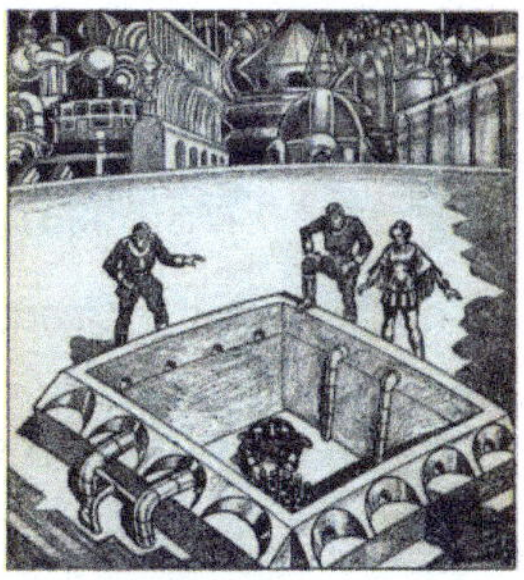

Part 4 (the conclusion) of *The Cometeers* by Jack Williamson appeared in *Astounding Stories* (date approximate).

Interior art by Marco Marchioni.

On this sol in 188 (1962 November 30):

During the third Kiwi-B nuclear thermal rocket engine test, a flame flash was observed when the reactor reached 120 MW. Power was

increased to 210 MW, and held there for 37 seconds. Power was then increased to 450 MW, but flashes then became frequent, and the engine was shut down after 13 seconds. After the test it was discovered that 97% of the fuel elements were broken. The difficulties of using liquid hydrogen were appreciated, and the cause of the vibration and failures was diagnosed as hydrogen leaking into the gap between the core and the pressure vessel. Unlike a chemical engine that would likely have blown up after suffering damage, the engine remained stable and controllable throughout. The tests demonstrated that a nuclear rocket engine would be rugged and reliable in space. This technology was being developed to provide propulsion for manned missions to Mars.

On this sol in 210 (2004 April 15):

MER-B Sol 81: *Opportunity* awoke with its alpha particle X-ray spectrometer on a soil target called Beagle Burrow inside a trench the rover had dug on Sol 73 (210 Dhanus 06). The rover removed the instrument arm, stowed it, then backed up to image the trench before driving toward a crater nicknamed Fram Crater. *Opportunity* then completed a 7.5-meter (24.6-foot) drive to a trough to image a rock outcrop within it with the panoramic camera. After a bit of guided driving, the rover set out using its autonomous navigation. The Sol 81 drive totaled more than 40 meters (131 feet).

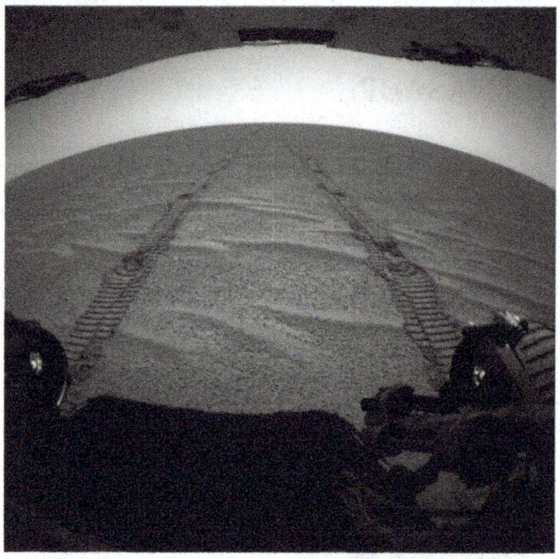

Right Rear Hazard Camera non-linearized full frame EDR acquired on Sol 81 of *Opportunity's* mission to Meridiani Planum at approximately 13:25:02

Nearly reaching the second of four waypoints on the way to Fram Crater, the rover imaged its new surroundings to identify any future driving hazards. An afternoon nap preceded Sol 81's final science session, atmospheric observations with the miniature thermal emission spectrometer and the panoramic camera.

—NASA/JPL

221 Dhanus 15 **2024 December 25 00:18 UTC**

Ls 020.6° Curiosity Sol 4403
MSD 53672 Perseverance Sol 1368

The first sol of the second quarter of Spring on the Aitken calendar

On this sol in 143 (1878 April 11):

Holger-Madsen was born; Danish film director, actor, and screenwriter; directed *Himmelskibet (Sky Ship)* (1918).

On this sol in 210 (2004 April 16):

MER-B Sol 82: Three days after switching to new software with mobility-enhancing features, NASA's *Opportunity* shattered the record for a single day's driving on Mars. The rover covered 140.9 meters (462 feet) during its 82nd sol on Mars. That is about 40 meters farther than either the best previous one-day drive, by *Opportunity* two weeks ago, or the total distance covered by NASA's smaller *Sojourner* rover during its entire three-month mission in 1997.

The first 55 meters (180 feet) was done as a "blind" guided drive based on images acquired previously. Speed during that session averaged 120 meters (394 feet) per hour. For the rest, *Opportunity* used autonomous navigation, watching for obstacles, choosing its own path, and averaging 40 meters (131 feet) per hour. After the drive, the rover took forward-looking images for planning the next drive.

Rover controllers devoted Sol 82 to driving after some morning atmospheric observations and a quick look back with the panoramic camera. The record-setting run took three hours — a good time for a marathon. It brought *Opportunity* to within about 90 meters (295 feet) of Fram Crater. It also took *Opportunity* over the 600-meter threshold, a criterion that had been set for at least one of the Mars Exploration Rovers to achieve in order for the mission to be called a success. *Opportunity* has now traveled 627.7 meters (0.39 mile). *Spirit* passed the 600-meter threshold two weeks ago.

—NASA/JPL

On this sol in 210 (2004 April 17):

MER-A Sol 103: *Spirit* had a busy weekend, culminating with a 75-meter (246-feet) drive toward Missoula Crater. The sol before the drive, Sol 102 (210 Dhanus 15), was an easier sol for *Spirit.* Its main objectives were to use the panoramic camera and mini thermal emission spectrometer to acquire photometric and atmospheric measurements.

Before beginning the drive on Sol 103, *Spirit* took panoramic camera images to help planners localize the rover during the long traverse. It then used the panoramic camera and mini thermal emission spectrometer to take a look back at the wheel tracks. Once this information was onboard, the rover began to drive.

Rover controllers planned the first 37 meters (121.4 feet) of the drive, but *Spirit* used the updated autonomous navigation software to see it through the remaining 38 meters (124.7 feet). Between the two drives, *Spirit* imaged its surroundings with the panoramic and navigation cameras for context. At the end of the 75-meter (246-feet) drive, *Spirit* rested a mere 40 meters (131.2 feet) from its destination at the rim of Missoula Crater. From that spot, it took mini thermal emission spectrometer observations of the sky and ground along with panoramic and navigation camera images to plan the next drive.

—NASA/JPL

221 Dhanus 16 **2024 December 26 00:57 UTC**

Ls 021.1° Curiosity Sol 4404

MSD 53673 Perseverance Sol 1369

First sol of Quinquatria, celebrated on ante dies XIV Kalendas Aprilis on the Roman calendar (analogous to March 44 on the revised Pickering calendar); a celebration sacred to Mars and Minerva, the first of five sols. The Salii (priests of Mars) danced in the comitium (attended by the pontiffs and the symbolic representatives of the army—the tribuni celerum), and the sacred arma ancilia were purified. In this sense, it is a ritual preparation for the season's coming military campaigns.

On this sol in 141 (1874 July 11):

Walter Franklyn Barrett was born; Australian film director and cinematographer; wrote, produced, directed, and filmed the first known motion picture about Mars, "A Message from Mars" (1903), believed to be lost.

On this sol in 153 (1897 February 01):

Part 5 of *The Martian* by George du Maurier appeared in *Harper's New Monthly Magazine* (date approximate).

On this sol in 154 (1898 December 21):

Reginald Hubert Rogers was born; Canadian pulp artist; created the interior art for "The Treasure of Ptakuth" (1940) by Leigh Brackett.

On this sol in 210 (2004 April 18):

MER-A Sol 104: This was a remote sensing day for *Spirit.* It included a search for dust devils and panoramic camera imaging of Mars's moon Phobos as it transits across the Sun and sets.

—NASA/JPL

221 Dhanus 17 **2024 December 27 01:37 UTC**

L_s 021.6° Curiosity Sol 4405
MSD 53674 Perseverance Sol 1370

Second sol of Quinquatria, celebrated on ante dies XIII Kalendas Aprilis on the Roman calendar (analogous to March 45 on the revised Pickering calendar); a celebration sacred to Mars and Minerva.

On this sol in 158 (1906 July 01):

A *Gift from Mars* (Part 3 of 3) by William Wallace Cook appeared in *The All-Story Magazine* (date approximate).

On this sol in 187 (1961 January 15):

Pathfinders to Mars Episode 6 "Falling into the Sun" airdate.

On this sol in 210 (2004 April 18):

MER-B Sol 84: *Opportunity* began the sol with some remote sensing observations and analysis with the microscopic imager and Mössbauer spectrometer. At 13:13 Mars Local Solar time, *Opportunity* began a 25-meter (82 feet) drive toward Fram Crater, taking images of its surroundings on the way. At the conclusion of the drive the rover acquired more remote sensing.

Opportunity will spend Sol 85 using the instruments on its instrument deployment device to investigate a rock target at Fram Crater.

On this sol in 210 (2004 April 19):

MER-A Sol 105: This mosaic was assembled from images taken by the panoramic camera on the Mars Exploration Rover *Spirit* at a region dubbed Site 31. *Spirit* is looking at Missoula Crater. From orbit, the features within the crater appeared to be ejecta from the younger Bonneville Crater, but *Spirit*'s closer look revealed wind-blown drift deposits, not ejecta, within Missoula Crater.

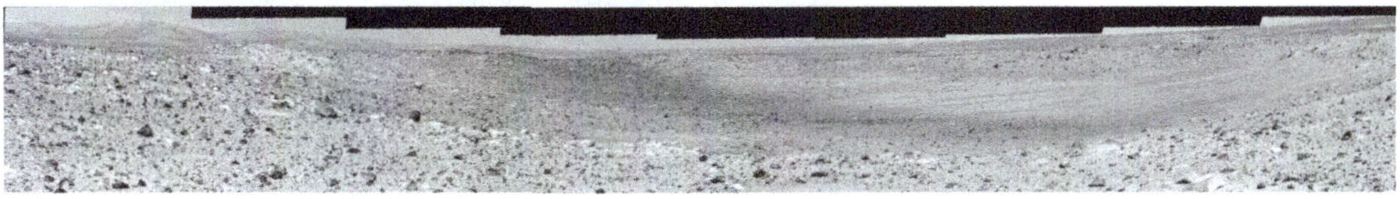

221 Dhanus 18 **2024 December 28 02:16 UTC**

Ls 022.0° Curiosity Sol 4406

MSD 53675 Perseverance Sol 1371

Third sol of Quinquatria, celebrated on ante dies XIII Kalendas Aprilis on the Roman calendar (analogous to March 46 on the revised Pickering calendar); a celebration sacred to Mars and Minerva.

On this sol in 210 (2004 April 20):

MER-B Sol 85: *Opportunity* got an up-close look at the rocky nature of Fram Crater as it approached the crater. After some morning remote sensing, the rover drove to the target rock dubbed "Pilbara," near the crater rim.

Plans called for *Opportunity* to grind into Pilbara with its rock abrasion tool on Sol 86.

On this sol in 210 (2004 April 20):

MER-A Sol 106: *Spirit* spent most of the sol performing remote sensing on the inside of Missoula Crater. It acquired panoramic camera and mini thermal emission spectrometer panoramas and navigation camera images of the crater, along with some panoramic camera images looking back toward Bonneville Crater.

—NASA/JPL

221 Dhanus 19 **2024 December 29 02:56 UTC**

Ls 022.5° Curiosity Sol 4407

MSD 53676 Perseverance Sol 1372

Fourth sol of Quinquatria, celebrated on ante dies XIII Kalendas Aprilis on the Roman calendar (analogous to March 47 on the revised Pickering calendar); a celebration sacred to Mars and Minerva.

On this sol in 171 (1930 December 15):

"The Mark of the Meteor" by Ray Cummings appeared in *Wonder Stories Quarterly*. Three human beings isolated in the immensity of space—death from starvation and suffocation nears. Who will weaken first!

Cover and interior art by Frank R. Paul.

"The Martian Nemesis" by George B. Beattie also appeared in *Wonder Stories Quarterly*. Man seeking to conquer strange worlds must put himself in the same position as a Martian on reaching the earth. Every plant, every living thing would be foreign to him. He might eat the poisonous fruits because they looked good, he would be stung by snakes, attacked by all manner of wild beasts. The more abundant the plant and animal life the greater the chance of his coming to grief and destruction.

Interior art by Marco Marchioni.

On this sol in 210 (2004 April 20):

MER-B Sol 86: *Opportunity's* 86th sol was another record-breaker A nearly two-and-a-half hour grind produced an impressive 7.2 millimeter (about 0.28 inches) hole in the rock called Pilbara.

The plan for the rest of the sol called for placing the alpha particle X-ray spectrometer on the new impression to determine the elemental composition of the exposed area. It was determined, however, that the rover position would not allow for a safe integration of the instrument. Rover planners amended the plan for the sol so the rover would back up and reposition itself for a safe placement of the spectrometer after the rock abrasion tool completed the grind.

On Sol 87 the rover will analyze the rock abrasion tool hole with the alpha particle X-ray and Mössbauer spectrometers.

On this sol in 210 (2004 April 21):

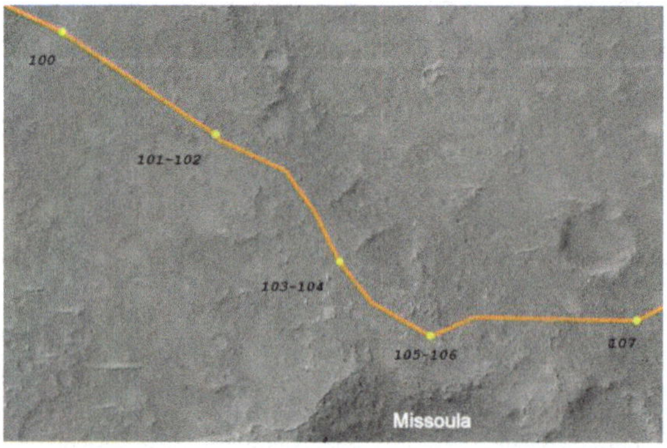

MER-A Sol 107: In the morning, *Spirit* got some atmospheric and cloud observations with the panoramic camera and mini thermal emission spectrometer. The rover then took a look with the panoramic camera at three targets called Gratteri Piazza, Wallula Gap, and Clark Fork. Finally, it was time to drive. *Spirit* completed a 73.8-meter (242-feet) traverse that included a jog around a sandy hollow to the east of Missoula. Most of the drive was in the southeast direction on course to the Columbia Hills.

After the drive, *Spirit* acquired additional panoramic camera and navigation camera observations. The total odometry at the end of Sol 107 was 976.77 meters (0.6 miles).

—NASA/JPL

221 Dhanus 20 **2024 December 30 03:36 UTC**

Ls 023.0° Curiosity Sol 4408

MSD 53677 Perseverance Sol 1373

Fifth sol of Quinquatria, celebrated on ante dies XIII Kalendas Aprilis on the Roman calendar (analogous to March 48 on the revised Pickering calendar); a celebration sacred to Mars and Minerva.

Mars Reconnaissance Orbiter is scheduled to begin its eleventh marsrev of orbital operations.

On this sol in 139 (1870 October 07):

Marcianus Filomeno Rossi was born; Italian science fiction writer; author of *A Trip to Mars* (1920). A crew of explorers set out to visit Mars on a solar powered space craft. When they arrive, they find the planet is apparently home to the descendants of an ancient Roman civilization. Their captain is asked to give names to canals on the planet and chooses to name them for Schiaparelli (credited with having first observed the alleged canals on Mars) and Flammarion (a popularizer of the idea of life on other worlds). The book includes this rather detailed drawing of a spaceship, the *Aeriolus*. The ship is described as having a locked hull similar to a space ship, and comes with a set of oxygen tanks so the crew can breathe during the flight. Interestingly, it uses a helicopter to ascend into space and then transitions to using the kick back from a set of guns to create a propelling force in the vacuum of space.

—Scribd.com

On this sol in 183 (1953 July 11):

Space Patrol, Season 3 Episode 28, "The Mysterious Ocean in Space" airdate. Commander Corry and his crew discover a massive body of water flowing in the middle of space. Investigating, they learn that criminals on Mars are in the process of stealing of Earth's water unless the planet's government pays them an extraordinary ransom.

—International Movie Database

On this sol in 210 (2004 April 21):

MER-B Sol 87: Opportunity spent its 87th sol gathering compositional information from the depression ground into Pilbara on Sol 86. The Mössbauer spectrometer examined the hole before the alpha particle X-ray spectrometer was placed there. The microscopic imager shot close-ups of Pilbara's new impression.

Data were also gathered by the miniature thermal emission spectrometer.

This set of activities should nearly complete a very detailed look at representative rocks and soil from Fram Crater, which can then be compared to the Eagle Crater rocks and soils.

MER-A Sol 108: *Spirit* continued driving toward the Columbia Hills.

—NASA/JPL

221 Dhanus 21 **2024 December 31 04:15 UTC**
L_S 023.5° Curiosity Sol 4409
MSD 53678 Perseverance Sol 1374

On this sol in 210 (2004 April 22):

MER-B Sol 88: This view in approximately true color reveals details in an impact crater informally named Fram in the Meridian Planum region of Mars. The picture is a mosaic of frames taken by the panoramic camera on NASA's Mars Exploration Rover *Opportunity* during the rover's 88th sol. The crater spans about 8 meters (26 feet) in diameter. *Opportunity* paused beside it while traveling from the rover's landing site toward a larger crater farther east. This view combines images taken using three of the camera's filters for different wavelengths of light: 750 nanometers, 530 nanometers and 430 nanometers.

MER-A Sol 109: *Spirit* continued driving toward the Columbia Hills.

—NASA/JPL

221 Dhanus 22 **2025 January 01 04:55 UTC**
L_S 024.0° Curiosity Sol 4410
MSD 53679 Perseverance Sol 1375

On this sol in 156 (1902 October 01):

John Tucker Battle was born; American screenwriter; wrote the story for the motion picture *Invaders from Mars* (1953).

On this sol in 210 (2004 April 24):

MER-B Sol 89: *Opportunity's* microscopic imager photographed a soil target called Nougat within the scuff. A Mössbauer spectrometer reading of the target followed.

The photometry experiment continued on this sol along with miniature thermal emission spectrometer remote sensing.

MER-A Sol 110: *Spirit* began the with a stretch of its "arm" to take microscopic imager pictures of an area of soil called Waffle Flats. It then placed the Mössbauer spectrometer instrument on that spot for a 90-minute integration. *Spirit* did double-duty and was able to get panoramic camera and mini thermal emission spectrometer images of the area for localization and science purposes while the Mössbauer was at work.

Spirit then stowed its instrument deployment device and began an 80-meter (262.5 feet) drive, half of it directed by rover planners and half using the autonomous navigation software. During the autonomous navigation portion, the rover detected a hazard and did not complete the final short-drive intended at the end of the journey. Images from the front hazard avoidance camera show no sign of a hazard, leaving rover controllers with a bit of a mystery to investigate.

Following the drive, *Spirit* took panoramic camera and navigation camera images in the drive direction and performed atmospheric science with the panoramic camera and mini thermal emission spectrometer.

—NASA/JPL

221 Dhanus 23 **2025 January 02 05:34 UTC**

Ls 024.4° Curiosity Sol 4411
MSD 53680 Perseverance Sol 1376

On this sol in 210 (2004 April 25):

MER-B Sol 90: With 811.57 meters (a little over one half of one mile) on its odometer, 12,429 images downloaded and a record for the longest one-sol drive under its belt, *Opportunity* completed its prime mission. Finishing 90 sols of surface operations since landing day marked completion of the last of the official success criteria for *Opportunity's* prime mission.

The rover continued with the multi-sol panoramic camera and miniature thermal emission plains photometry observations. The alpha particle X-ray spectrometer was busy collecting data on the soil targets Nougat and Fred Ripple.

MER-A Sol 111: After a successful weekend of driving on Sols 108 and 109 (210 Dhanus 20 and 21), *Spirit* kicked off its week with a 140-meter (459.3 feet) drive over Sols 110 and 111 (210 Dhanus 22 and 23) toward its destination at the base of the Columbia Hills.

Sol 111 was also a sol full of driving for *Spirit.* After acquiring panoramic camera images of its surroundings and completing atmospheric science with the panoramic camera and mini thermal emission spectrometer, the rover began its drive.

Spirit successfully completed a 60.8-meter (199.5 feet) drive toward the Columbia Hills and then acquired navigation and panoramic camera images of the driving direction. *Spirit* ended the day with mini thermal emission spectrometer observations of the soil and then a coordinated mini thermal emission spectrometer and panoramic camera study of the atmosphere.

—NASA/JPL

221 Dhanus 24 **2025 January 03 06:14 UTC**

Ls 024.9° Curiosity Sol 4412

MSD 53681 Perseverance Sol 1377

On this sol in 210 (2004 April 25):

MER-B Sol 91: *Opportunity's* extended mission commenced. *Opportunity* completed more remote sensing and took a Mössbauer spectrometer read on Fred Ripple. The rest of the sol was spent driving. A 40-meter (131.2 feet) drive in the southeasterly direction left Opportunity only 160 meters (about 525 feet) from the rim of Endurance Crater.

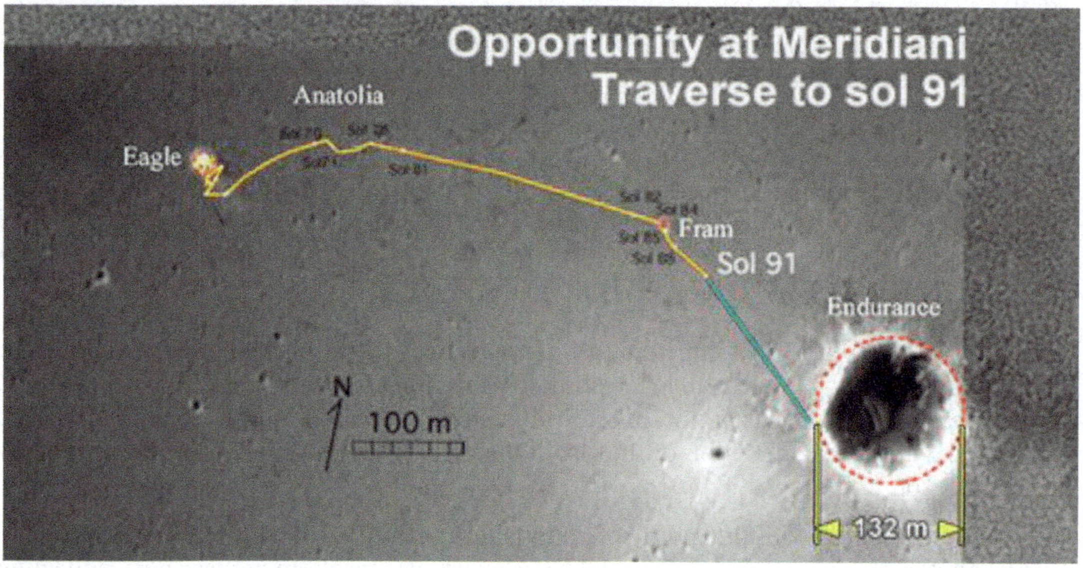

On this sol in 210 (2004 April 26):

MER-A Sol 112: *Spirit* took it easy in the morning and did not begin operations until 11:45 a.m. to conserve energy for an afternoon drive. Before taking off, *Spirit* gathered some soil and atmospheric observations with the mini thermal emission spectrometer and panoramic camera.

Then the drive began. *Spirit's* updated autonomous navigation software proved its worth again this sol. During a long auto-navigation segment, the rover encountered a hazard and was able to back up and find a way around it. *Spirit* continued to drive backwards towards its intended goal point, using the rear hazard avoidance cameras to navigate the way. When the allotted drive time was up, *Spirit* turned back around and made one last short drive to its resting place for the night. *Spirit's* odometer records backwards and forwards driving and logged another 88.6 meters (290.7 feet) for the sol 112 drive. The actual distance covered was about 60 meters (197 feet).

—NASA/JPL

On this sol in 188 (1962 December 12):

US President John F. Kennedy decided to postpone the Reactor-In-Flight-Test (RIFT), a Saturn V-launched Nuclear Engine for Rocket Vehicle Application (NERVA) flight demonstration planned for 1967, until after additional Kiwi-B nuclear thermal rocket engine ground tests had occurred, explaining that "the nuclear rocket... would be useful for further trips to the moon or trips to Mars. But we have a good many areas competing for our available space dollars, and we have to channel it into those programs which will bring a result—first, our moon landing, and then consider Mars."

Participating in NASA Marshall Space Flight Center's Early Manned Planetary – Interplanetary Roundtrip Expedition (EMPIRE) study, Aeronutronic issued its report describing a 21-month flyby mission on a modified Crocco flight profile to be launched toward Mars during a 28-day window opening 1970 July 19. The study considered both chemical and thermal nuclear propulsion options.

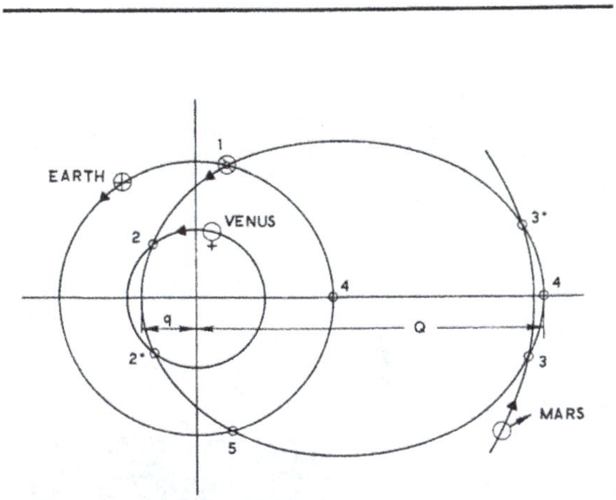

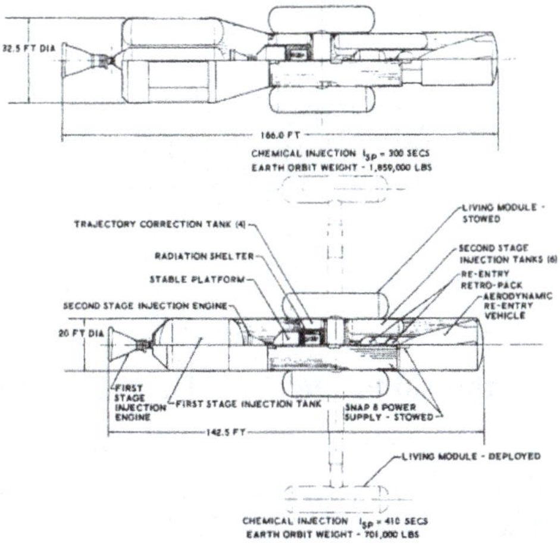

The reentry vehicle, in which the crew would have returned to Earth, was to have been a lifting body, a wingless vehicle which develops aerodynamic lift from the shape of its body. A manned lifting body vehicle has yet to be developed to this day, although a number of subscale designs have been flight-tested.

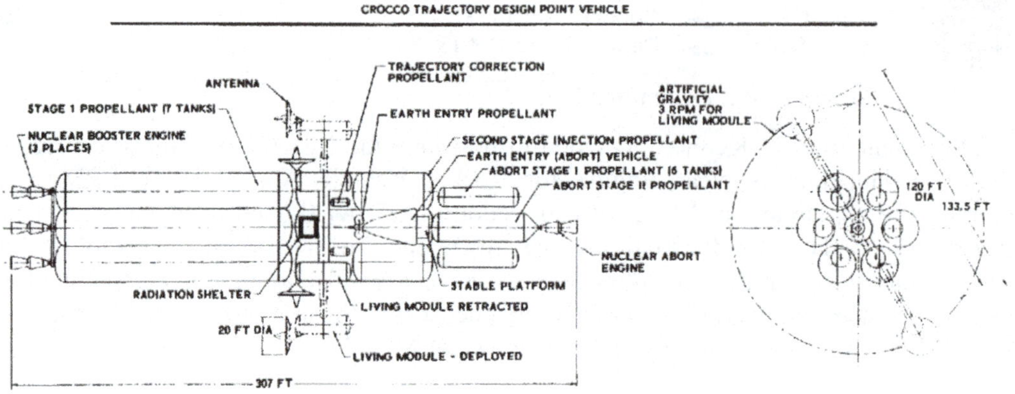

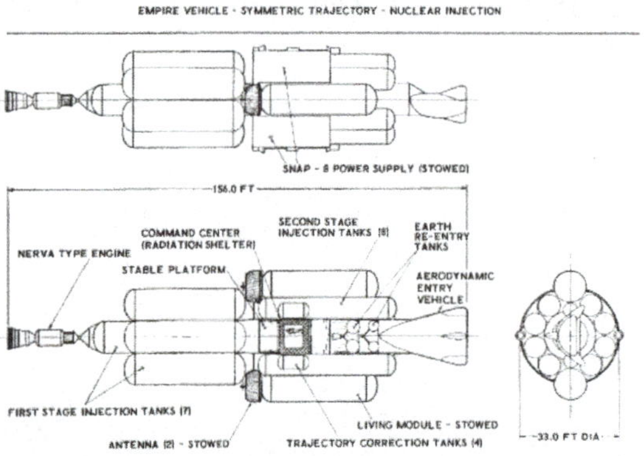

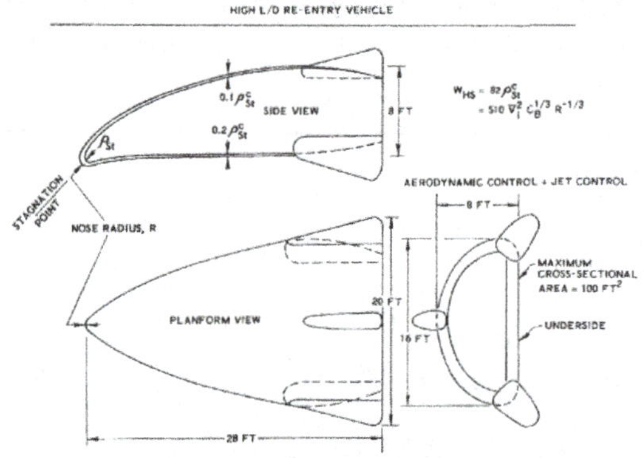

On this sol in 196 (1977 December 29):

A 150 million-year-old, 482.5-gram peridotite (lherzolitic shergottite) meteorite designated ALHA 77005 was discovered in the Allan Hills, Antarctica. In the 1980s, the meteorite was identified as having originated on Mars.

On this sol in 210 (2004 April 26):

MER-B Sol 92: Opportunity continued toward Endurance Crater.

On this sol in 210 (2004 April 27):

MER-A Sol 113: *Spirit* woke up earlier than normal, 9:00 a.m., to do morning atmospheric science. One objective of the early sky scan was to image morning clouds with the panoramic camera. *Spirit* then began an intense study of a soil spot called MayFly. During her examination of the area, *Spirit* took panoramic camera and mini thermal emission spectrometer images in parallel, conducted a two-hour Mössbauer integration and finished off with a look through the microscopic imager. The rover then stowed the instrument arm to prepare for digging a trench.

Rover planners intended for *Spirit* to use its wheels to dig a trench at the MayFly spot, but hazard avoidance camera images of the area showed a potato-size rock that could have potentially fallen into the wheel hollow in the process. Rather than take that risk, controllers decided to back the rover up 10 centimeters (3.9 inches) to a clearer spot. After the final positioning, *Spirit* used its wheels to dig a 6-centimeter (2.4-inch) trench. *Spirit* finished the sol with hazard avoidance camera images of the trench, which was used to plan Mössbauer, alpha particle X-ray spectrometer and microscopic imager work on Sol 114 (210 Dhanus 27).

—NASA/JPL

221 Dhanus 26 **2025 January 05 07:33 UTC**
Ls 025.9° Curiosity Sol 4414
MSD 53683 Perseverance Sol 1379

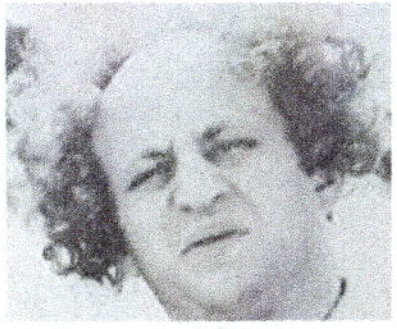

On this sol in 156 (1902 October 05):

Louis Feinberg (Larry Fine) was born; American actor and screenwriter; starred in the motion picture *The Three Stooges in Orbit* (1962).

On this sol in 175 (1938 July 01):

"The Secret of the Canali" by Clifton B. Kruse appeared in *Astounding Science Fiction* (date approximate). Two surveyors mapping Mars's ancient secret—the Canali—stumble on the Door through which the Martians fled.

Interior art by Elliott Dold, Jr.

"Rule 18" by Clifford D. Simak also appeared in *Astounding Science Fiction* (date approximate). A rule defeated Earth teams in the annual Earth-Mars football game—till a coach pulled the prize bright trick of—quite literally—all time.

Interior art by Charles Schneeman.

On this sol in 210 (2004 April 27):

MER-B Sol 93: *Opportunity* edged its way closer to Endurance Crater. A total drive of 106 meters (347.8 feet) on Sols 91 and 92 (210 Dhanus 24 and 25) left the rover just 70 meters (229.7 feet) from the rim.

The pattern for these two sols has been to take pre- and post-drive remote sensing observations and imaging in the crater direction between midday energy-conserving naps.

By Sol 95 (210 Dhanus 28), Opportunity will make the final approach to Endurance Crater.

On this sol in 210 (2004 April 28):

MER-A Sol 114: *Spirit* continued to investigate the trenched area with the Mössbauer spectrometer, alpha particle X-ray spectrometer and the microscopic imager.

Spirit performed a lot of science activities in the trench called Big Hole using the microscopic imager, Mössbauer spectrometer and alpha particle X-ray spectrometer. *Opportunity* also studied the rover tracks and the crater rim.

—NASA/JPL

221 Dhanus 27 **2025 January 06 08:13 UTC**
Ls 026.4° Curiosity Sol 4415
MSD 53684 Perseverance Sol 1380

Aṅgārakadina (अङ्गारकदिन), a festival of Angaraka (Mars) on the Hindu lunar calendar, on the fourteenth of the latter half of the month of Chaitra (चैत्र) (analogous to March 55 on the revised Pickering calendar).

On this sol in 210 (2004 April 28):

MER-B Sol 94: *Opportunity* drove 50-meters (164-feet).

—NASA/JPL

221 Dhanus 28 **2025 January 07 08:52 UTC**
Ls 026.8° Curiosity Sol 4416
MSD 53685 Perseverance Sol 1381

As seen from Mars, the Sun enters the constellation Capricornus.

On this sol in 210 (2004 April 29):

MER-B Sol 95: *Opportunity* drove the final approach of 17 meters (56 feet) to arrive on the western rim of Endurance Crater and began surveying the spectacular new view.

Opportunity sits about half a meter (1.6 feet) outside the edge of the crater with a positive pitch of 4.7 degrees, meaning the rover is slightly tilted with its head up. The western side of the crater rim slopes down in front of *Opportunity* with an angle of about 18 degrees for about 17 meters (56 feet).

On this sol in 210 (2004 April 30):

MER-A Sol 116: The sol started with a repeat of the microscopic imaging of a target in the trench due to minor communication glitches on Sol 115. *Spirit* then stowed the arm, backed away from Big Hole trench, and took panoramic camera images of the trench before it continued on its trek toward the Columbia Hills. The drive on Sol 116 established a new drive record of 90.8 meters (298 feet) for *Spirit.*

—NASA/JPL

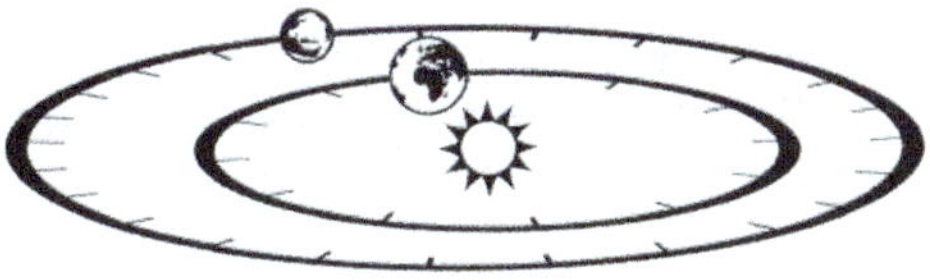

221 CAPRICORNUS

As seen from Mars, the Sun traverses the western sector of the constellation Capricornus, which has been consistently represented as a hybrid of a goat and a fish since the Middle Bronze Age (3,500 to 3,200 years before present).

The median one-way communications delay during this month is 5 minutes 26 seconds.

← 221 Capricornus 01 **2025 January 08 09:32 UTC →**

Ls 027.3° Curiosity Sol 4417
MSD 53686 Perseverance Sol 1382

Approximately the first sol of October on the Heinlein calendar.

On this sol in 210 (2004 April 30):

MER-B Sol 96: *Opportunity* operators focused on remote sensing of Endurance Crater and the interesting features in and around it.

On this sol in 210 (2004 May 01):

MER-A Sol 117: *Spirit* drove 37 meters (121 feet) to a small ridge, where the vehicle experienced a pitch up of 12.2 degrees. Engineers believe that the change in tilt caused the vehicle to recompute its "goodness map," which helps the rover autonomously drive over the Martian terrain, and the rover declared that it was not safe to continue its drive. One good

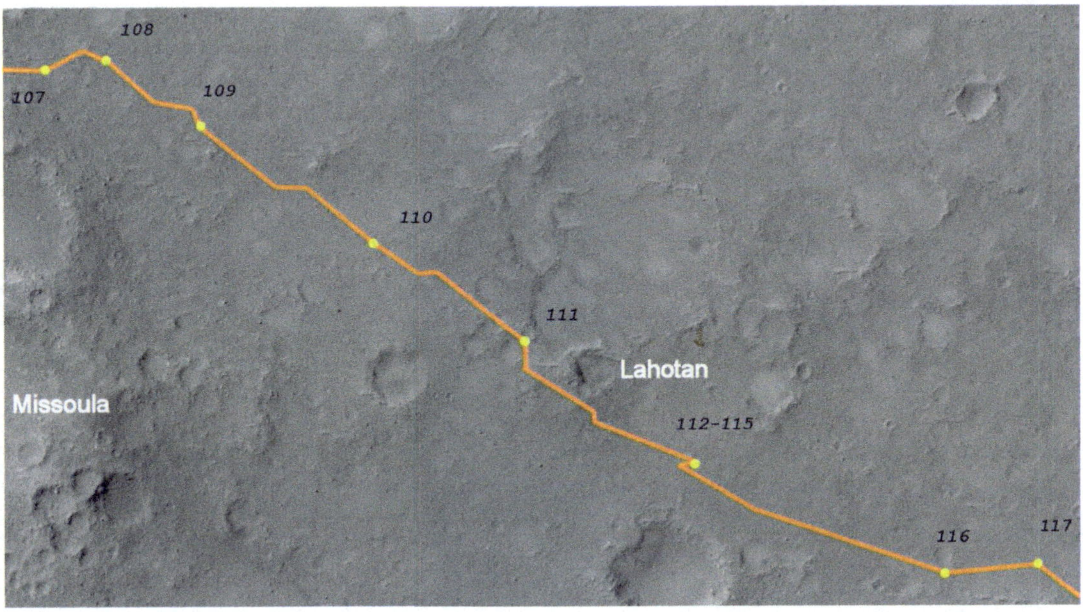

thing that came out of this is that the end-of-drive tilt positioned the solar arrays to maximize afternoon solar exposure, and the rover's battery state of charge is in good health.

—NASA/JPL

221 Capricornus 02 **2025 January 09 10:11 UTC**
Ls 027.8° Curiosity Sol 4418
MSD 53687 Perseverance Sol 1383

On this sol in 210 (2004 May 01):

MER-B Sol 97: *Opportunity* operators focused on remote sensing of Endurance Crater and the interesting features in and around it.

On this sol in 210 (2004 May 02):

MER-A Sol 118: The Gusev Crater rover *Spirit* moved 92.4 meters (303 feet) across the surface in one sol, breaking its previous record of around 90 meters (295 feet). The *Opportunity* rover still has *Spirit* beat with a one-sol driving record of 140 meters (459.3 feet).

—NASA/JPL

221 Capricornus 03 **2025 January 10 10:51 UTC**
Ls 028.3° Curiosity Sol 4419
MSD 53688 Perseverance Sol 1384

On this sol in 210 (2004 May 02):

MER-B Sol 98: *Opportunity* continued to gaze at Endurance Crater from its vantage point on the western rim. Remote sensing included gathering imagery of two potential traverse targets just inside the northern edge and southwestern edge of the crater.

On this sol in 210 (2004 May 03):

MER-A Sol 119: An uplink configuration error prevented the sequence load from successfully getting on board the rover *Spirit.* Rover controllers took advantage of the down day by deleting afternoon communication sessions and enabling the rover to charge its battery during a long afternoon nap.

—NASA/JPL

221 Capricornus 04 **2025 January 11 11:31 UTC**
Ls 028.7° Curiosity Sol 4420
MSD 53689 Perseverance Sol 1385

On this sol in 210 (2004 May 03):

MER-B Sol 99: *Opportunity* continued to gaze at Endurance Crater from its vantage point on the western rim. Remote sensing included gathering imagery of two potential traverse targets just inside the northern edge and southwestern edge of the crater.

On this sol in 210 (2004 May 04):

MER-A Sol 120: *Spirit* is now approximately 1.7 kilometers (1 mile) away from the base of the Columbia Hills after three long sols of driving. Its odometer currently reads 1,566 meters (0.97 miles) and counting.

This cylindrical-projection 120-degree image mosaic was created from three navigation camera images that NASA's Mars Exploration Rover *Spirit* acquired on Sol 120. The image highlights a crater approximately 70 meters (230 feet) in diameter that scientists have informally named Lahontan. This image also reveals a wind-ripple feature in the foreground and a distant look at the Columbia Hills on the horizon, *Spirit's* planned final destination.

—NASA/JPL

221 Capricornus 05 **2025 January 12 12:10 UTC**
Ls 029.2° Curiosity Sol 4421
MSD 53690 Perseverance Sol 1386

On this sol in 210 (2004 May 05):

MER-A Sol 121: *Spirit* conducted atmospheric measurements before continuing its trek toward the Columbia Hills. A 96.8-meter (318 feet) drive that consisted of about half direct drive and half auto-navigational drive broke *Spirit's* last one-sol distance traveled. That drive brought the mission total to 1,669 meters (1.04 miles), flipping the rover's odometer over the one-mile mark.

—NASA/JPL

221 Capricornus 06 **2025 January 13 12:50 UTC**
Ls 029.7° Curiosity Sol 4422
MSD 53691 Perseverance Sol 1387

On this sol in 141 (1874 July 27):

Frank Connolly Shannon was born; American actor; starred in the serial motion picture *Flash Gordon's Trip to Mars* (1938).

On this sol in 210 (2004 May 06):

MER-A Sol 122: It was a touch-and-go sol for *Spirit,* starting with a half-hour alpha particle X-ray spectrometer integration, a one-hour Mössbauer integration and a set of four microscopic images all on the same patch of soil. Panoramic camera and miniature thermal emission spectrometer data were also obtained before an afternoon nap. The bulk of the afternoon was spent driving another 65 meters (213 feet).

—NASA/JPL

221 Capricornus 07 **2025 January 14 13:29 UTC**
Ls 030.2° Curiosity Sol 4423
MSD 53692 Perseverance Sol 1388

On this sol in 150 (1891 July 01):

Otis Adelbert Kline was born; American songwriter and science fiction writer; author of “The Man from the Moon” (1930), *The Swordsman of Mars* and *Outlaws of Mars* (both 1933).

On this sol in 210 (2004 May 07):

MER-B Sol 102: *Opportunity* awoke from its first “deep sleep.” This set of activities was initiated to conserve the energy that is being used by the instrument arm’s stuck-on heater switch. During deep sleep, rover planners power off the main electronics at night and open the switches that supply battery power to the main power bus, and in turn nearly all the secondary electronics. In particular, this removes power input to the Rover Power Distribution Unit, which normally supplies power to the stuck-on heater. With the Rover Power Distribution Unit input turned off, the heater cannot burn any energy either. In the morning, when the sun strikes the solar panel array, the Battery Control Board resets and connects the batteries to the main power bus again. At this time, the stuck-on heater again draws power, but this will only be for a few hours in the morning instead of all night.

The most vulnerable instrument to the cold Martian nights is the miniature thermal emission spectrometer. With a cutoff of the power electronics, its heater cannot keep it warm overnight. Data returned on Sol 102 showed the temperature reached -46 degrees Celsius (-50.8 degrees Fahrenheit), a bit warmer than the spectrometer’s lowest proven temperature for functionality, -50 degrees Celsius (-58 degrees Fahrenheit).

Rover planners commanded *Opportunity* to take a drive during the afternoon of Sol 102 to the south, along the edge of the crater toward a dark rock in the vicinity.

More remote sensing was conducted, including miniature thermal emission spectrometer measurements that confirmed the instrument is still functioning normally after deep sleep.

MER-A Sol 123: The sol started off with *Spirit’s* panoramic camera and miniature thermal emission spectrometer observations for near-field surveys, atmospheric studies, and localization. *Spirit* then took a half-hour nap, followed by the day’s drive. This sol consisted of another 48-meter (about 157-feet) direct drive, the mid-drive survey and localization remote sensing, and then 47-meters (about 154 feet) of driving using auto-navigation. The total was 95.2 meters (312 feet), bringing the mission total to 1830 meters (1.14 miles).

—NASA/JPL

221 Capricornus 08 **2025 January 15 14:09 UTC**
Ls 030.6° Curiosity Sol 4424
MSD 53693 Perseverance Sol 1389

Opposition occurs at 12:09 Airy Mean Time (2025 January 16 02:38 UTC).

On this sol in 210 (2004 May 08):

MER-B Sol 103: *Opportunity* traversed approximately 13 meters (about 43 feet) farther south along the eastern rim of Endurance Crater, reaching the beginning of the Karatepe area.

MER-A Sol 124: *Spirit* drove 80 meters (262.5 feet) on Sol 124, bringing its total odometry to 1,909.52 meters (1.2 miles). *Spirit* has less than 1.2 kilometers (.75 mile) to go before reaching the base of the Columbia Hills and will reach them by Sol 160. Later in the sol, after completing the Sol 124 drive, *Spirit* took a 360-degree afternoon panorama of its surroundings with the navigation camera.

—NASA/JPL

221 Capricornus 09 **2025 January 16 14:49 UTC**
Ls 031.1° Curiosity Sol 4425
MSD 53694 Perseverance Sol 1390

The first sol of May, marsrev 3679 on the Levitt calendar.

On this sol in 150 (1891 July 04):

René Gontran Ranson was born; French science fiction cover artist; created the cover art for *Le mystère des XV: Le triomphe de l'amour (The Mystery of the XV: The Triumph of Love)* (1922).

On this sol in 153 (1897 February 23):

James Morgan Walsh was born; Australian science fiction writer; author of *Vandals of the Void* (1931) and *Terror Out of Space* (1934).

On this sol in 210 (2004 May 09):

MER-B Sol 104: *Opportunity* approached Lion Stone, a rock at Endurance Crater's edge that stands about 10 centimeters tall (about 4 inches) and is about 30 centimeters long (12 inches). This brought *Opportunity's* total mission odometry to 1,054 meters (3,458 feet).

MER-A Sol 125: *Spirit* continued driving and set a new one-sol driving record of 123.7 meters (405.8 feet). Science on Sol 125 included morning atmospheric sky and ground remote sensing, mini thermal emission spectrometer observation of the Sol 126 instrument deployment device work volume, imaging with the panoramic camera, and cloud observations.

—NASA/JPL

221 Capricornus 10 **2025 January 17 15:28 UTC**
Ls 031.6° Curiosity Sol 4426
MSD 53695 Perseverance Sol 1391

On this sol in 210 (2004 May 10):

MER-B Sol 105: *Opportunity* acquired a series of microscopic images of Lion Stone and the surrounding soil.

The rover then went on to collect a short Mössbauer integration on the rock during the day, performed a tool change to the alpha particle X-ray spectrometer in late afternoon.

MER-A Sol 126: After the long Sol 125 drive, *Spirit* was in perfect position to work with the instrument deployment device on Sol 126. This included alpha particle X-ray spectrometer, Mössbauer and microscopic imager work on a target called Lead Foot (in honor of the big drive on Sol 125). The Mössbauer was used as the feeler for all these activities but touched down on rocks rather than soil at the Lead Foot location, compromising the Mössbauer and microscopic imager data (images out of focus). *Spirit* also did some driving on this sol and added 55.6 meters (182.4 feet) to the odometer, bringing *Spirit's* new drive total to 2,089 meters (1.3 miles). At the end of the sol, *Spirit* successfully executed a sequence that used the panoramic camera to find the Sun and correct for accumulated rover attitude errors.

—NASA/JPL

---

221 Capricornus 11 **2025 January 18 16:08 UTC**
Ls 032.0° Curiosity Sol 4427
MSD 53696 Perseverance Sol 1392

On this sol in 210 (2004 May 11):

MER-B Sol 106: *Opportunity* acquired the alpha particle X-ray spectrometer integration in the early morning. That sol also included additional microscopic images and a successful "bump" maneuver to reposition the rover so the top of Lion Stone was in position for the rock abrasion tool on Sol 107. Remote sensing was also acquired during Sols 105 and 106, including panoramic camera images of the heatshield that protected *Opportunity* during its toasty trip through the Martian atmosphere. The heatshield impacted approximately 250 meters (about 820 feet) south of Endurance Crater.

MER-A Sol 127: *Spirit* spent most of the sol continuing its drive toward the Columbia Hills. The rover put approximately 70 more meters (229.7 feet) on its odometer and then took an hour-and-20-minute siesta. After the drive, *Spirit* took observations with the panoramic camera, navigation camera and mini thermal emission spectrometer.

—NASA/JPL

---

221 Capricornus 12 **2025 January 19 16:47 UTC**
Ls 032.5° Curiosity Sol 4428
MSD 53697 Perseverance Sol 1393

On this sol in 176 (1940 June 01):

"Warlords of Mars" by Festus Pragnell appeared in *Amazing Stories* (date approximate). approximate). Don Hargreaves didn't want to fight the warlords of Mars, but he couldn't ignore a princess's promise.

Interior art by Robert Fuqua.

On this sol in 191 (1968 August 18):

David Crosby's song "Triad" was included as a track on the American rock band Jefferson Airplane's vinyl LP *Crown of Creation,* a year after it was recorded by Crosby's own group, The Byrds (see 211 Mithuna 18). Although not specifically about Mars, Crosby's lyrics, and indeed the free love movement of the 1960s, were influenced by Robert A. Heinlein's novel *Stranger in a Strange Land* (1961), with specific references to "sister lovers" and "water brothers," thus it may be thought of as the first Martian love song.

You want to know how it will be
Me and him or you and me
You both stand there, your long hair flowing
Your eyes alive, your mind still growing
Saying to me - what can we do now that we both love you
I love you too, I don't really see
Why can't we go on as three

You are afraid, embarrassed too
No one has ever said such a thing to you
Your mother's ghost stands at your shoulder
Face like ice, a little bit colder
Saying to you - you can not do that, it breaks
All the rules you learned in school
I don't really see
Why can't we go on as three

We love each other, it's plain to see
There's just one answer comes to me
Sister lovers, water brothers
And in time maybe others
So you see what we can do is to try something new
If you're crazy too, I don't really see
Why can't we go on as three

On this sol in 210 (2004 May 12):

MER-B Sol 107: *Opportunity* successfully drilled a hole into Lion Stone with the rock abrasion tool. Since the surface of the rock was fairly uneven, the tool had to work through some high spots before getting a good bite on the rock for a full circular hole. Sol 107 ended with a nighttime integration of the alpha particle X-ray spectrometer to reveal elemental composition of the inner part of the rock.

MER-A Sol 128: *Spirit* began the sol by completing a panoramic camera observation of a rock target called Flat Head. the rover then rested up for a couple of hours before embarking on a 90-meter (295 feet) drive toward the hills. Once the drive was complete, *Spirit* completed its standard post-drive observations.

—NASA/JPL

221 Capricornus 13 **2025 January 20 17:27 UTC**
Ls 033.0° Curiosity Sol 4429
MSD 53698 Perseverance Sol 1394

On this sol in 210 (2004 May 13):

MER-B Sol 108: The *Opportunity* rover finished up its work on Lion Stone by analyzing the rock abrasion tool hole with the Mössbauer spectrometer and taking microscopic images to create a mosaic of the hole. *Opportunity* then moved away from Lion Stone and continued traversing counterclockwise around the crater. *Opportunity* drove 32 meters (105 feet) to the top of a small ridge for a better view of where to drive on Sol 109.

MER-A Sol 129: The sol began successfully, but *Spirit* encountered a couple of difficulties before the sol was over. After waking, *Spirit* performed 45 minutes of science observations and then settled down for a morning nap. With plenty of energy stored, it was time to drive. *Spirit* roved 31 meters (102 feet) across the surface in an engineer-directed drive and then spent 45 minutes using its autonomous navigation system to try to drive down the side of a small ridge. The backside slope of the ridge was too steep, and the autonomous navigation system had *Spirit* turn in an attempt to find another way down. Unfortunately, a couple of large rocks close to the ridge prevented *Spirit* from finding a safe path down. At the end of the drive sequence, *Spirit* was supposed to complete a "stutter step" to get in proper position to do work with the instrument deployment device on Sol 130. Unfortunately, the rover was unable to complete this final positioning or the ultimate post-drive imaging, so Sol 130 was mostly a drive sol.

—NASA/JPL

221 Capricornus 14 **2025 January 21 18:06 UTC**
Ls 033.4° Curiosity Sol 4430
MSD 53699 Perseverance Sol 1395

On this sol in 210 (2004 May 14):

MER-B Sol 109: *Opportunity* drove about 41 meters (135 feet).

—NASA/JPL

221 Capricornus 15 **2025 January 22 18:46 UTC**
Ls 033.9° Curiosity Sol 4431
MSD 53700 Perseverance Sol 1396

On this sol in 153 (1897 March 01):

Part 6 of *The Martian* by George du Maurier appeared in *Harper's New Monthly Magazine* (date approximate).

On this sol in 210 (2004 May 15):

MER-B Sol 110: *Opportunity* drove about 41 meters (135 feet). *Opportunity* ended the drive with a "scuff" of the soil and rocks under the front wheel. This scuff action produced an interesting dislodged plate of some kind.

Opportunity is driving along the south edge of Endurance Crater, with a southward tilt of about 8 degrees. The Sun is now at higher latitudes (south hemisphere winter is coming), so a southward tilt robs the rover of total solar array energy. This is making it more difficult to perform many activities. In a couple of sols when *Opportunity* drives to flatter ground near the crater edge to take the next large panorama, the energy situation is expected to improve.

MER-A Sol 131: *Spirit* took an unplanned break on its trek to the Columbia Hills due to a software fault.

Spirit used its panoramic camera to take the images that make up this full-resolution mosaic of the Columbia Hills. *Spirit* acquired these images on Sol 131, when it sat less than one kilometer (0.62 miles) away from its target, the base of the hills. *Spirit* will reach the base before Sol 160 (210 Aquarius 09) and begin an in-depth exploration of the interesting geologic features that make up these hills.

—NASA/JPL

221 Capricornus 16 **2025 January 23 19:26 UTC**
Ls 034.4° Curiosity Sol 4432
MSD 53701 Perseverance Sol 1397

On this sol in 174 (1936 September 01):

"A Beast of the Void" by Raymond Z. Gallun appeared in *Astounding Stories* (date approximate). The strange story of Lothar Weiss, a semi-recluse in a small Midwestern town, who stumbles upon a meteorite which was imbedded in a black, leathery-looking substance. Cutting away the meteoritic metal with an oxyacetylene torch, he observes that the mysterious object is seemingly unaffected by heat. Removed, it turns out to be a flattish disk with claws around the edge, much like the top of a small umbrella. After a time, Weiss discovers that the object is alive....

—Bleiler and Bleiler, 1998

Interior art by C.R. Thomson.

On this sol in 210 (2004 May 16):

MER-B Sol 111: *Curiosity* drove on a 40-meter (131 feet) traverse along the southern edge of Endurance Crater. drive put *Opportunity* on an 8-degree slope that tilted the rover away from the Sun and limited the amount of direct sunlight that could reach the solar panels.

MER-A Sol 132: *Spirit* spent another sol on break during its trek to the Columbia Hills due to a software fault on Sol 131 (210 Capricornus 15). That fault left rover planners with some uncertainty about *Spirit's* final position and attitude, so Sol 132 was spent re-establishing that knowledge with panoramic, navigation and hazard avoidance camera imaging of the rover's surroundings.

—NASA/JPL

221 Capricornus 17 **2025 January 24 20:05 UTC**
Ls 034.8° Curiosity Sol 4433
MSD 53702 Perseverance Sol 1398

On this sol in 210 (2004 May 17):

MER-B Sol 112: *Opportunity* is healthy, but feeling a bit sluggish today. The previous sol's 40-meter (131 feet) traverse along the southern edge of Endurance Crater, and a Sol 112 error with a Deep Space Network command transmission have resulted in a low battery state of charge.

To help the battery recover to its normal state of charge, rover planners had built a Sol 112 plan that deleted two of the three UHF windows. Unfortunately, a Deep Space Network configuration error prevented the command load from reaching *Opportunity* on Sol 112 and, as expected in such cases, the rover executed the onboard run-out sequence, which included an hour of remote sensing and the three on-board UHF communication windows.

MER-A Sol 133: It appears that *Spirit's* software error on Sol 131 (210 Capricornus 15) occurred within a 3-microsecond window of vulnerability when a "write" command was attempted to a "write-protected" area of RAM. The flight software team believes this is an extremely low probability event, and has not adjusted the planning process to avoid the miniscule period of vulnerability. *Opportunity* has the same vulnerability to the fault.

The unplanned break in *Spirit* operations did have a silver lining, though; it resulted in fully charged batteries, paving the way for a long drive on Sol 133. *Spirit* roved 113 meters (370.7 feet) on Sol 133, with a record 78-meter (256 feet) autonomous navigation segment. The previous record for an autonomous navigation drive was 62 meters (203.4 feet) on Sol 125 (210 Capricornus 09). *Spirit's* odometer now reads 2473 meters (1.53 miles) and it is roughly 780 meters (0.5 miles) from the Columbia Hills and in excellent health.

—NASA/JPL

221 Capricornus 18 **2025 January 25 20:45 UTC**
Ls 035.3° Curiosity Sol 4434
MSD 53703 Perseverance Sol 1399

On this sol in 149 (1889 August 24):

Leonard Thomas Clapham (Tom London) was born; American actor; starred in the motion picture *Nan of the North : Missil from Mars* (1922).

On this sol in 210 (2004 May 18):

MER-B Sol 113: *Opportunity* slept and recharged.

MER-A Sol 134: *Spirit* roved an impressive 109.5 meters (359.3 feet) on this sol. Two hours of the drive were guided by the autonomous navigation system. After the long traverse, *Spirit* completed an hour of post-drive science observations with the panoramic and navigation cameras and mini thermal emission spectrometer. The rover finished the sol healthy and ready for another sol on Mars.

—NASA/JPL

221 Capricornus 19 **2025 January 26 21:24 UTC**
Ls 035.8° Curiosity Sol 4435
MSD 53704 Perseverance Sol 1400

On this sol in 183 (1953 August 08):

Rod Brown of the Rocket Rangers, Season 1 Episode 17, "The Martian Queen" airdate.

On this sol in 210 (2004 May 19):

MER-B Sol 114: Planned activities for *Opportunity* included some limited remote sensing in the morning, with most of the sol spent recharging the batteries. The limited activity over the next few sols will focus on moving towards the Endurance Crater rim and a new position for panoramic camera imaging.

MER-A Sol 135: After so much driving on Sol 134 (210 Capricornus 18), *Spirit* got a break and spent Sol 135 doing in-situ science investigations of its surroundings. It began the sol observing nearby soil with the alpha particle X-ray spectrometer and Mössbauer spectrometer. It then used the microscopic imager to see the soil up close. After stowing the instrument deployment device, *Spirit* used its wheels to dig a trench and then imaged the trench with the cameras on the mast.

—NASA/JPL

221 Capricornus 20 — 2025 January 27 22:04 UTC

Ls 036.2° — Curiosity Sol 4436
MSD 53705 — Perseverance Sol 1401

On this sol in 156 (1902 October 28):

Elsa Sullivan Lanchester was born; British-American actress; starred in the motion picture *Pajama Party* (also titled *The Maid and the Martian)* (1964).

On this sol in 172 (1932 December 01):

The Wreck of the Asteroid by Laurence Manning began serialization in *Wonder Stories* (date approximate). A sequel to "The Voyage of the Asteroid." Think of being stranded on Mars, faced with every imaginable kind of enemy. What can the ingenuity of man conceive to effect a rescue? Will he weakly resign himself to his fate, eventual death, of fight for life with all the weapons at his command?

Interior art by Frank R. Paul.

"Roadways of Mars" by Harl Vincent appeared in *Amazing Stories* (date approximate). Swerving sharply from the main roadway along Canal Pyramus, the glistening the nickel-cobalt surface of a narrow, seldom-used branch stretched out into the desert.

Interior art by Leo Morey.

On this sol in 210 (2004 May 20):

MER-B Sol 115: *Opportunity* drove 11.7 meters (38.4 feet), coming to rest about 3 meters (10 feet) from the edge of Endurance Crater, as intended. Rover planners had commanded *Opportunity* to go 10 centimeters (3.9 inches) farther, but the rover decided to stop when it saw the edge of the crater in the navigation camera images. This was actually a more conservative response than necessary, as it would have been safe to complete the drive. Rover planners are looking into changing the way they send commands to prevent this over-conservatism next time.

MER-A Sol 136: Engineers sent *Spirit* its commands and received the beep confirming that they were running. However, the afternoon pass by NASA's *Mars Odyssey* orbiter did not provide any data, and the orbiter reported that it had not heard from *Spirit's* UHF antenna. Engineers first thought was that the signal might have been blocked by the Columbia Hills because of the low elevation pass of *Mars Odyssey*. The telecommunications team disagreed with this hypothesis and thought something might be wrong. Scientists and engineers would have to wait until the afternoon of Sol 137 (210 Capricornus 21) for their next communications opportunity.

—NASA/JPL

221 Capricornus 21 **2025 January 28 22:44 UTC**

Ls 036.7° Curiosity Sol 4437

MSD 53706 Perseverance Sol 1402

On this sol in 210 (2004 May 21):

MER-B Sol 116: *Opportunity* used its navigation camera to acquire images showing its proximity to the crater. The rover turned slightly to the right and crept a little closer to the edge of Endurance Crater to get into just the right position to set up camp for a few sols. The rover executed this 1.5-meter (4.9 feet) traverse as planned, ending up facing northwest with a total tilt of about 8 degrees pitched nose-up. From this position, *Opportunity* will make many observations with the panoramic camera and mini thermal emission spectrometer to fully characterize the parts of the crater that can be seen from here. *Opportunity* now sits only about 1 meter from the edge of the crater, and there is a sloping drop-off of about 40 degrees dead ahead.

MER-A Sol 137: During Sols 136 and 137, engineers executed a number of communication trials with *Spirit,* but did not receive any telemetry from the rover, or have any indication of what had gone wrong until finally data came back from a Sol 137 afternoon pass by NASA's *Mars Global Surveyor.* That pass revealed that on Sol 136, *Spirit's* software had experienced an extremely low probability error that rebooted the computer and terminated all the sequences. This error created a domino effect of communication difficulties and explained why engineers had not been able to make contact with the rover. All the anomaly events are understood.

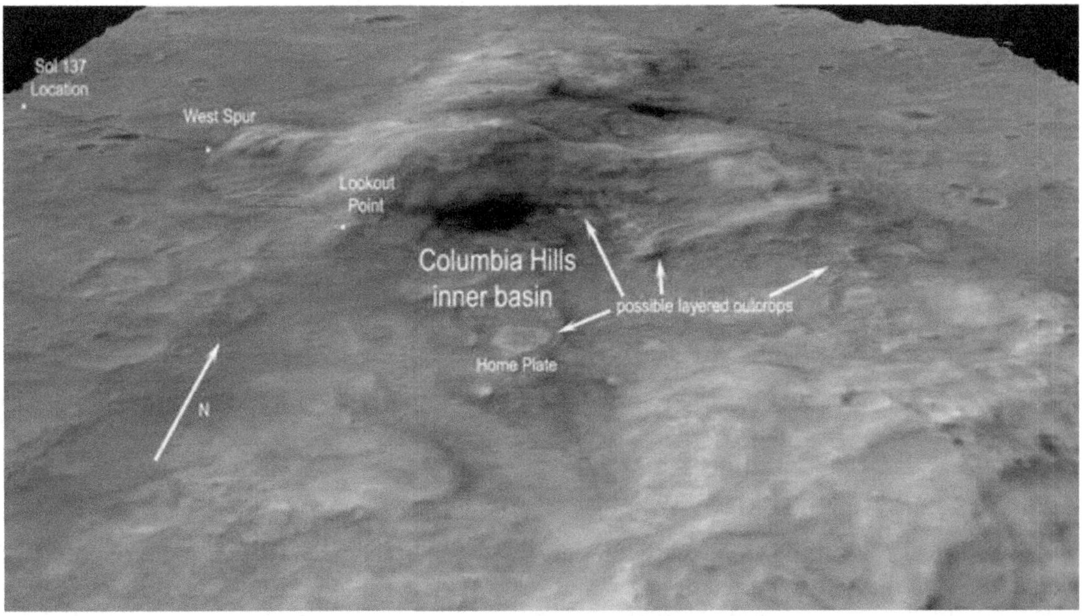

This view of the Columbia Hills in Gusev Crater was made by draping an image from the Mars Orbiter Camera on NASA's *Mars Global Surveyor* orbiter (image E0300012 from that camera) over a digital elevation model that was derived from two Mars Orbiter Camera images (E0300012 and R0200357).

This unique view is helpful to the rover team members as they plan the journey of NASA's Mars Exploration Rover *Spirit* to the base of the Columbia Hills and beyond. *Spirit* successfully completed a three-month primary mission, and so far remains healthy in an extended mission of bonus exploration. As of Sol 135 (221 Capricornus 19, 2004 May 21), *Spirit* sits approximately 680 meters (0.4 miles) away from its first target at the western base of the hills, a spot informally called West Spur. The team estimates that *Spirit* will reach West Spur by Sol 146 (210 Makara 02, 2004 June 01). *Spirit* will most likely remain there for about a week to study the outcrops and rocks associated with this location.

When done there, *Spirit* will head approximately 620 meters (0.38 miles) to a higher-elevation location informally called Lookout Point. *Spirit* might reach Lookout Point by around Sol 165 (210 Makara 21, 2004 June 20). On the way, the rover will pass by and study ripple-shaped wind deposits that may reveal more information about wind processes on Mars.

Lookout Point will provide a great vantage point for scientists to remotely study the inner basin area of the Columbia Hills. This basin contains a broad range of interesting geological targets including the informally named "Home Plate" and other possible layered outcrops. These features suggest that the hills contain rock layers. *Spirit* might investigate the layers to determine whether they are water-deposited sedimentary rock.

Once at Lookout Point, *Spirit* will acquire 360-degree panoramic images of the entire area to help define the rover's next steps. Assuming the rover stays healthy, *Spirit* will eventually drive down into the basin to get an up-close look at interesting features there.

—NASA/JPL

221 Capricornus 22 **2025 January 29 23:23 UTC**

Ls 037.2° Curiosity Sol 4438
MSD 53707 Perseverance Sol 1403

On this sol in 210 (2004 May 22):

MER-A Sol 138: The sol was spent recovering the state of the rover *Spirit,* including reestablishing a master sequence, restoring high-gain antenna communications, and reinitializing the panoramic camera mast actuators whose positions had been marked as unknown.

—NASA/JPL

221 Capricornus 23 **2025 January 31 00:03 UTC**

Ls 037.6° Curiosity Sol 4439
MSD 53708 Perseverance Sol 1404

On this sol in 210 (2004 May 24):

MER-A Sol 139: Engineers have developed a way to reduce even further the probability of encountering this particular error again and reported that on the bright side, *Spirit* was able to fully charge its batteries during the anomaly.

Spirit had plans to spend Sols 136 through 139 (210 Capricornus 20 through 23) observing its surroundings with the remote sensing instruments on its mast, and then exploring the trench it dug on Sol 135 (210 Capricornus 19) with the instruments on the robotic arm. However, an anomaly on Sol 136 restricted the rover's activity, allowing *Spirit* to achieve only remote science objectives during the four-sol segment. *Spirit* has since fully recovered and has resumed normal science operations.

Spirit performed the remote-sensing observations that had been lost along with some observations coordinated with *Mars Global Surveyor* that had been planned for Sol 138 (210 Capricornus 22). All activities on Sol 139 completed successfully, which verified that *Spirit* had returned to normal science operations.

—NASA/JPL

221 Capricornus 24 **2025 February 01 00:42 UTC**

Ls 038.1° Curiosity Sol 4440
MSD 53709 Perseverance Sol 1405

On this sol in 210 (2004 May 25):

MER-A Sol 140: *Spirit* Sol 140 picked up with the original plan for Sol 137 (210 Capricornus 21), performing in-situ work in the trench.

Spirit spent most of Sol 140 investigating the trench it dug on Sol 135 (210 Capricornus 19). It got an up-close look at the trench with the microscopic imager and then began a five-and-a-half-hour integration with the Mössbauer spectrometer. Doing double duty, *Spirit* surveyed the Columbia Hills with the panoramic camera and miniature thermal emission spectrometer during the Mössbauer observation. Following the afternoon *Mars Odyssey* communications window, *Spirit* changed tools and began an overnight integration on two targets with the alpha particle X-ray spectrometer.

—NASA/JPL

221 Capricornus 25 **2025 February 02 01:22 UTC**
Ls 038.6° Curiosity Sol 4441
MSD 53710 Perseverance Sol 1406

On this sol in 210 (2004 May 26):

MER-A Sol 141: *Spirit* continued to observe the trench with the Mössbauer spectrometer and microscopic imager. The rover also obtained panoramic camera images of its surroundings while doing work with the robotic arm. It then stowed the instrument deployment device and backed up 0.85 meters (2.8 feet). *Spirit* spent most of the afternoon observing the trench using the miniature thermal emission spectrometer.

—NASA/JPL

221 Capricornus 26 **2025 February 03 02:02 UTC**
Ls 039.0° Curiosity Sol 4442
MSD 53711 Perseverance Sol 1407

On this sol in 210 (2004 May 27):

MER-A Sol 142: *Spirit* got a chance to stretch its wheels and do some driving on Sol 142, but before taking off, the rover finished the trench observations with some panoramic camera imaging. Then it was time to rove. *Spirit* completed a 30-minute, engineer-directed drive and then turned the wheel over to the autonomous navigation software for another hour and 15 minutes of driving. *Spirit* roved a total of 61 meters (200.1 feet) closer toward the Columbia Hills.

—NASA/JPL

221 Capricornus 27 **2025 February 04 02:41 UTC**
Ls 039.5° Curiosity Sol 4443
MSD 53712 Perseverance Sol 1408

On this sol in 183 (1953 August 15):

Rod Brown of the Rocket Rangers, Season 1 Episode 18, "The Fire Demons of Deimos" airdate.

On this sol in 210 (2004 May 28):

MER-A Sol 143: *Spirit* began Sol 143 using the panoramic camera to image its surroundings. After a restful nap, the rover began driving and advanced 69 meters (226.4 feet) in an engineer-directed drive toward the Columbia Hills. After doing some mid-drive science imaging, *Spirit* attempted additional driving using its autonomous navigation software, but detected hazards in the immediate vicinity of the rover and therefore covered no additional ground.

—NASA/JPL

221 Capricornus 28 **2025 February 05 03:21 UTC**
Ls 040.0° Curiosity Sol 4444
MSD 53713 Perseverance Sol 1409

Earth departure opportunity to Mars via an EVM trajectory.

Mars arrival opportunity from Earth via an EEM trajectory.

Earth arrival opportunity from Mars via a Type I trajectory.

On this sol in 210 (2004 May 29):

MER-A Sol 144: *Spirit* covered 24.4 meters (80 feet) in another engineer-directed drive. Once again, the terrain was too rough to permit further driving beyond this point using the autonomous navigation software.

—NASA/JPL

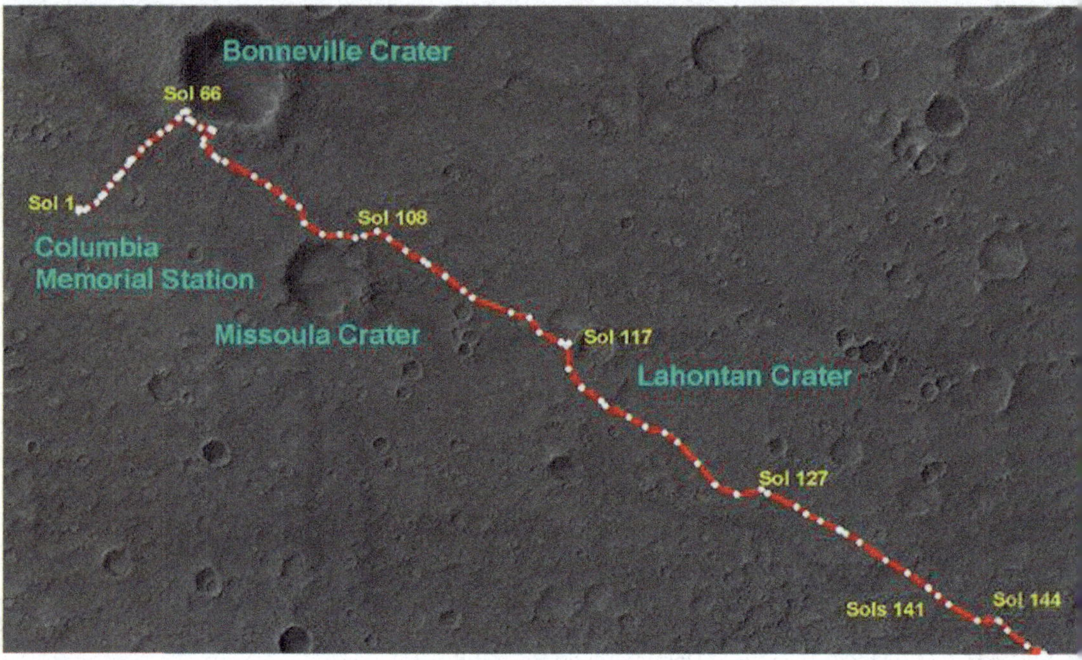

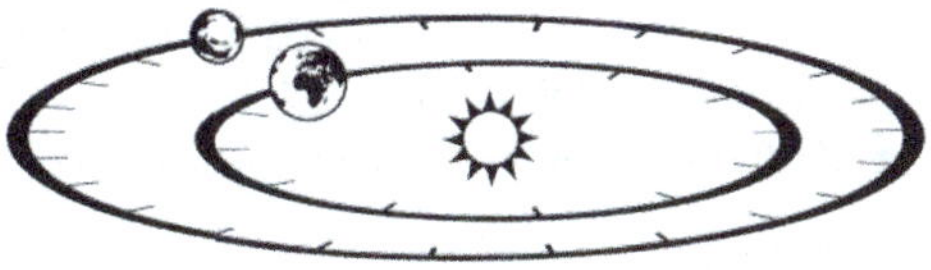

221 MAKARA

As seen from Mars, the Sun traverses the eastern sector of the constellation Capricornus. The name Makara (मकर), the Sanskrit word for "sea dragon" or "water-monster," is taken from a month in the Hindu solar calendar that originally occurred when the Sun passed through the constellation Capricornus as seen from Earth.

The median one-way communications delay during this month is 6 minutes 38 seconds.

221 Makara 01 **2025 February 06 04:00 UTC**

Ls 040.4° Curiosity Sol 4445
MSD 53714 Perseverance Sol 1410

The first sol of the third quarter of Spring on the Aitken calendar; approximately the first sol of the month Zeus on the Heinlein calendar.

On this sol in 140 (1872 October 01):

James Allen St. John was born; American author, artist and illustrator; created the cover art for *The Warlord of Mars* (1919), *Thuvia, Maid of Mars* (1920), *The Chessmen of Mars* (1922), *The Master Mind of Mars* (1928), *Swords of Mars* (1936), "Invisible Men of Mars" (1941), "John Carter and the Giant of Mars" (1941), and "Yellow Men of Mars" (1941) by Edgar Rice Burroughs.

On this sol in 175 (1938 August 01):

"The Thing from Mars" by Ray Cummings appeared in *Thrilling Wonder Stories* (date approximate). Laboratory science produces an incredible form of matter!

Interior art by Leo Morey.

On this sol in 184 (1955 July 07):

The *X Minus One* presentation of the radio drama "The Green Hills of Earth" by Robert A. Heinlein was broadcast on KFI in Los Angeles. A song-writing space engineer is blinded in an accident; later, when nearing the end of his life, he wishes to make a final trip home to see the "green hills of earth." The broadcast included the poet Rhysling's "The Grand Canal" (see 220 Taurus 09).

On this sol in 210 (2004 May 30):

MER-A Sol 145: *Spirit* completed a 43-meter (141 feet) engineer-directed drive and then spent two hours roving another 55 meters (180 feet) using the autonomous navigation software.

—NASA/JPL

221 Makara 02 **2025 February 07 04:40 UTC**

Ls 040.9° Curiosity Sol 4446
MSD 53715 Perseverance Sol 1411

On this sol in 185 (1957 May 24):

Adventures of Superman, Season 5 Episode 12, "Mr. Zero" airdate. *The Daily Planet* gets a strange visitor who turns out to be a rejected citizen from Mars. It is discovered Mr. Zero has a talent that attracts the attention of some local gangsters who draw Mr. Zero into their gang.

—International Movie Database

On this sol in 210 (2004 May 31):

MER-A Sol 146: *Spirit* roved 61 meters (200 feet) on Sol 146.

—NASA/JPL

221 Makara 03 **2025 February 08 05:19 UTC**

Ls 041.4° Curiosity Sol 4447
MSD 5376 Perseverance Sol 1412

On this sol in 152 (1895 May 01):

Part 1 of *Mars* by Percival Lowell appeared in *The Atlantic* (date approximate).

On this sol in 210 (2004 May 31):

MER-B Sol 126: *Opportunity* began the sol with a quick placement of the Mössbauer instrument on Diogenes. The rover then napped for about two hours while the Mössbauer performed its integration. Upon waking, the rover stowed its arm and began a mobility test and preparation activity that will aid rover planners should they decide to traverse down steep rocky slopes in Endurance Crater. This set of activities included a draw-bar pull activity where the front rover wheels are locked and dragged back across rocks by driving the other four wheels backwards for about one meter (3.3 feet). The draw-bar pull is intended to give insight into the friction between the wheels and the rock surface at this site. The other mobility preparation activity was to scuff each wheel on the surface by driving one wheel at a time for a few rotations in each direction (with all the other wheels locked). This "pawing at the ground" activity was intended to scrub off the anodized layer on the surfaces of the wheels, which will allow for better grip. After these mobility activities, *Opportunity* traversed about 72 meters (236.2 feet) west around the crater towards Karatepe. This very busy day also included relaying about 190 Megabits of data through *Mars Odyssey* with two back-to-back afternoon passes. All this was made possible by having the option of deep sleep to save energy overnight, which *Opportunity* took full advantage of overnight into Sol 127.

The plan for the coming sols involves approaching the rim of Endurance and imaging potential entry points to aid in the decision of whether or not to enter the crater.

MER-A Sol 147: *Spirit* roved 52 meters (171 feet) on Sol 147. At its current rate, *Spirit* is on schedule for a Sol 160 arrival at the base of the Columbia Hills.

—NASA/JPL

---

221 Makara 04 **2025 February 09 05:59 UTC**

Ls 041.8° Curiosity Sol 4448

MSD 53717 Perseverance Sol 1413

On this sol in 210 (2004 June 01):

MER-B Sol 127: After a "deep sleep" overnight, *Opportunity* began its 127th sol with a three and one-half-hour nap. Upon awaking, the rover drove 50 meters (164 feet) on a directed drive, turned to face Endurance Crater and took images using its front hazard-avoidance cameras. *Opportunity* then performed 45 minutes of remote sensing using the panoramic camera and miniature thermal emission spectrometer. The rover then supported an afternoon and overnight *Mars Odyssey* data pass.

On this sol in 210 (2004 June 02):

MER-A Sol 148: This was a driving sol, with *Spirit* completing a 64.7-meter (212.3 feet) engineer-directed drive. This put the rover in position for some Sol 149 (210 Makara 05) work with the robotic arm, and provided a great view of the Columbia Hills.

—NASA/JPL

---

221 Makara 05 **2025 February 10 06:39 UTC**

Ls 042.3° Curiosity Sol 4449

MSD 53718 Perseverance Sol 1414

On this sol in 173 (1934 November 01):

"Valley of Dreams" by Stanley G. Weinbaum appeared in *Wonder Stories* (date approximate). Sequel to "A Martian Odyssey." The further adventures on Mars with Tweel, the unforgettable bird-like alien, and the deadly, insidious dream-beasts that trap and kill by illusion.

Interior art by Frank R. Paul.

"The Martian Cry" by Lloyd Arthur also appeared in *Wonder Stories* (date approximate).

Red dust enfolds our world,
    A crimson blight
That fills our empty seas with arid Death.
In dull despair we sink, through wells of night
So deep, so dry, we. choke and gasp for breath.
Unending thirst flows flaming through, our veins,
Destroying life and hope,
    Yet hope lives on,
And raises croaking prayers for cooling rains
To lave our burning sphere till drought is gone.

We see a green and youthful world in space,
The lusty Earth,
With water it can give
To check the doom that menaces our race,
And help the dying planet, Mars, to live.
Canals and. seas are thick with dust, and dry.
We must have water,
Earthmen, or we die!

On this sol in 187 (1961 March 01):

The animated short film "The Kid from Mars" was released (date approximate). A little Martian lands on Earth near a circus and, with his ray gun, wreaks havoc for the ringmaster. He causes trouble until the ringmaster offers him a job.

—Big Cartoon Database

On this sol in 210 (2004 June 02):

MER-B Sol 128: Foregoing deep sleep over the Sol 127 night, *Opportunity* spent Sol 128 approaching the rim of Endurance Crater. The drive took about 10 minutes and moved *Opportunity* 13.4 meters (44 feet), coming to rest about 5 meters (16.4 feet) from the lip of the crater. The remainder of the sol was spent supporting two afternoon *Mars Odyssey* passes. The rover then underwent its sixth deep sleep cycle into the morning of Sol 129.

On this sol in 210 (2004 June 03):

MER-A Sol 149: *Spirit* took a break from driving and surveyed the Columbia Hills with the panoramic camera and mini thermal emissions spectrometer. After that, the rover attained an alpha particle X-ray spectrometer observation of the filter magnet and capture magnet. *Spirit* takes a look at its magnets every now and then to assess what magnetically susceptible materials have accumulated. The last magnet check was on Sol 92 (210 Dhanus 04).

—NASA/JPL

221 Makara 06 **2025 February 11 07:18 UTC**

Ls 042.7° Curiosity Sol 4450
MSD 53719 Perseverance Sol 1415

On this sol in 210 (2004 June 03):

MER-B Sol 129: It was decided that *Opportunity* was not in the most advantageous location for entering the crater. The rover was commanded to move approximately 50 meters (164 feet) back along the crater rim, close to the rock called Lion Stone. After a 4.5-meter (14.7 feet) bump toward the rim and some navigation and hazard-avoidance camera images, the rover backed up 4.5 meters (14.8 feet) and then drove 8 meters (26.2 feet) toward Lion Stone. Power limited the total drive duration. Instead of utilizing the deep sleep mode overnight into Sol 130, *Opportunity* again supported two *Mars Odyssey* passes that returned a large volume of data to Earth.

After Sol 129, Opportunity's odometer read 1,395.91 meters (4,579.76 feet).

On this sol in 210 (2004 June 04):

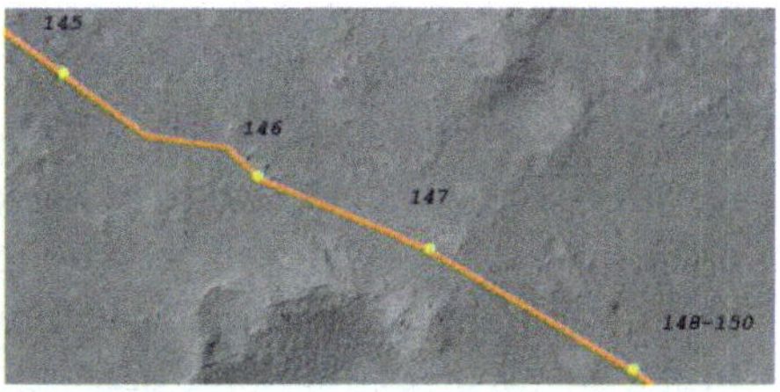

MER-A Sol 150: *Spirit* used its alpha particle X-ray spectrometer and Mössbauer spectrometer to observe a rock called Joshua. Unfortunately, the rest of the sol's planned work with the instrument deployment device did not take place because of a command anomaly, which made *Spirit* think that a collision between the rock abrasion tool and the forearm might occur. Therefore, the tool change and all subsequent arm motions were prevented for the rest of the sol.

—NASA/JPL

221 Makara 07 **2025 February 12 07:58 UTC**
Ls 043.2° Curiosity Sol 4451
MSD 53720 Perseverance Sol 1416

On this sol in 152 (1895 May 05):

Charles Lamont was born; American director, screenwriter, and producer; directed the motion picture *Abbott and Costello Go to Mars* (1953).

On this sol in 210 (2004 June 04):

MER-B Sol 130: *Opportunity* traversed a total of about 45 meters (about 147.6 feet). About 39 meters (about 127.9 feet) of that was counterclockwise along the edge of Endurance Crater, and 6 meters (about 19.7 feet) toward the crater rim. The sol ended with the rover about 10 meters (32.8 feet) from the crater rim. The traverse ended up about 1 meter (about 3.3 feet) short of what was commanded due to a slightly uneven patch of ground that the rover seemed to run across near the end of the drive. Driving over this tripped a suspension limit that rover planners had set to help prevent inadvertently driving into difficult terrain. Deep sleep was again invoked for the night of Sol 130 to 131.

On this sol in 210 (2004 June 05):

MER-A Sol 151: During Sols 148 through 151, *Spirit* advanced significantly closer to the Columbia Hills and now sits only 220 meters (722 feet) from its first target at the base, a location informally named Spur B.

Spirit was back to business on Sol 151 and finished observing Joshua and the science magnets with the tools on the instrument deployment device. After that, the rover was off, and successfully completed a 73-meter (240 feet) drive toward the Columbia Hills.

—NASA/JPL

221 Makara 08 **2025 February 13 08:37 UTC**
Ls 043.6° Curiosity Sol 4452
MSD 53721 Perseverance Sol 1417

Possible meteor shower associated with parent body P/2006 HR30 Siding Spring: Martian right ascension 240.9°, declination 63.3°.

On this sol in 210 (2004 June 05):

MER-B Sol 131: *Opportunity* successfully traversed up the slope to the crater edge, took a detailed set of images and then backed off a little to optimize its orientation for the rover's communications passes. These images will aid in the project's assessment of traversing on the interior slopes of Endurance Crater in this vicinity. Deep sleep was not invoked on this night, in favor of relaying data to *Mars Odyssey* in the early morning on Sol 132 (210 Makara 09).

On this sol in 210 (2004 June 06):

MER-A Sol 152: *Spirit* continued its journey toward the Columbia Hills and completed an 83-meter (272 feet) drive that brought its total odometry to 3.2 kilometers (2 miles). After the drive, the rover completed some remote sensing that brought more details of the hills into view.

—NASA/JPL

221 Makara 09 **2025 February 14 09:17 UTC**
Ls 044.1° Curiosity Sol 4453
MSD 53722 Perseverance Sol 1418

On this sol in 152 (1895 May 07):

Barney A. Sarecky was born; American screenwriter; co-wrote the screenplay for the motion picture *The Purple Monster Strikes* (1945).

On this sol in 210 (2004 June 06):

MER-B Sol 132: *Opportunity* re-approached the crater rim at the location and orientation most advantageous for the "pre-dip" into the crater. This approach was designed to just crest the edge of the crater and leave the rover roughly level (with the front two wheels in the crater). The drive executed beautifully.

—NASA/JPL

221 Makara 10 **2025 February 15 09:57 UTC**
Ls 044.6° Curiosity Sol 4454
MSD 53723 Perseverance Sol 1419

On this sol in 210 (2004 June 07):

MER-B Sol 133: *Opportunity* executed the first real dip into Endurance Crater. The intent was to go far enough in that all wheels would be on the slope of the crater, and then come all the way back out, proving that the rover was capable of getting back out before going very deep. The other main objective was to gather information on the degree and nature of any slip that would be experienced while traversing the crater wall. The execution went extremely well, with slips and disturbance of the terrain well below acceptable levels, giving the team confidence that the rover is capable of going deeper. The engineering team will continue to characterize the variety of slopes and materials that *Opportunity* will encounter deeper in the crater.

On this sol in 210 (2004 June 08):

MER-A Sol 154: *Spirit* roved another 70 meters (230 feet) on sol 153, and 49 meters (161 feet). After the drive, *Spirit* attained a miniature thermal emission spectrometer scan of the hills that will help scientists identify what the hills are made of.

—NASA/JPL

221 Makara 11 **2025 February 16 10:36 UTC**

Ls 045.0° Curiosity Sol 4455

MSD 53724 Perseverance Sol 1420

Midpoint between the northward equinox and the northern solstice, a cross-quarter sol analogous to Bealtaine, Floralia, Maiouma, or May Day (May 01) on Earth.

On this sol in 210 (2004 June 08):

MER-B Sol 134: *Opportunity* is becoming accustomed to its new sloped home inside Endurance Crater. There are positives and negatives to the rover's new position and orientation. The solar array is oriented toward the northeast, which maximizes solar power in the morning and also warms the high gain antenna actuator faster, so heating is no longer required before the morning communications session. On the downside, the UHF communications sessions have degraded slightly at this orientation.

On Sol 134, *Opportunity* drove 3.9 meters (about 13 feet) into Endurance Crater, then backed up 1.4 meters (4.6 feet), remaining inside the crater. Drive slippage and vehicle tilt was as predicted by the engineering team. An hour's worth of remote sensing completed the sol. *Opportunity* then performed deep sleep overnight into the morning of Sol 135 (210 Makara 12).

On this sol in 210 (2004 June 09):

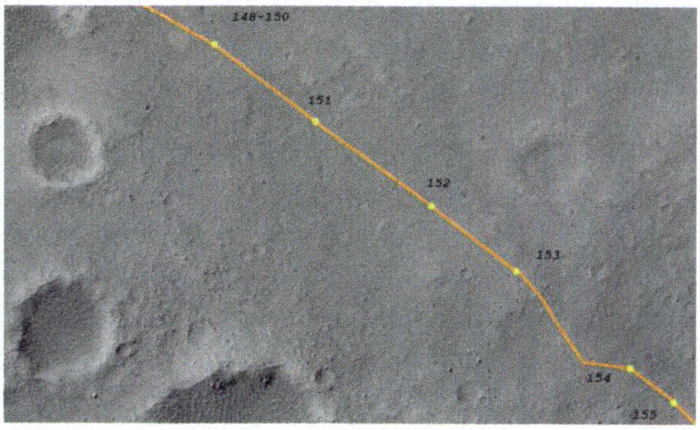

MER-A Sol 155: *Spirit* was roughly 50 meters (164 feet) from the base of the target location at the Columbia Hills. *Spirit* reached this location after a 23-meter (75 feet) drive that ended with the rover at a maximum tilt of 20 degrees. 20 degrees is well below the safe limit for tilt and was 3 to 4 degrees below the estimated tilt for this traverse.

—NASA/JPL

221 Makara 12 **2025 February 17 11:16 UTC**

Ls 045.5° Curiosity Sol 4456

MSD 53725 Perseverance Sol 1421

On this sol in 156 (1902 November 18):

Franklin Adreon was born; American director, producer, and screenwriter; directed the *Commando Cody: Sky Marshal of the Universe* episode "Robot Monster from Mars" (1955).

On this sol in 210 (2004 June 09):

MER-B Sol 135: *Opportunity* drove 1.3 meters (4.3 feet) deeper into Endurance Crater to a position that was about the deepest point it reached on sol 134. This short drive was intended to allow for detailed imaging of the first likely target for the instrument arm, a rock called Tennessee. The drive went exactly as planned, leaving *Opportunity* with a final tilt of -19.44 degrees and a heading of 62.5 degrees. The rover then performed almost two hours of remote sensing, then set up for another night of deep sleep.

On this sol in 210 (2004 June10):

MER-A Sol 156: *Spirit* roved 42 meters (138 feet) closer to a vantage point where it could observe the hill outcrops. Some of the images that *Spirit* sent back revealed a small and unusual rock that piqued scientists' interest and was informally named End-of-Rainbow.

—NASA/JPL

221 Makara 13 **2025 February 18 11:55 UTC**
Ls 045.9° Curiosity Sol 4457
MSD 53726 Perseverance Sol 1422

On this sol in 154 (1899 February 12):

Edward L. Cahn was born; American director; directed the motion picture *It! The Terror from Beyond Space* (1958).

On this sol in 210 (2004 June 11):

MER-B Sol 136: *Opportunity* spent the sol performing a series of panoramic camera and miniature thermal emission spectrometer observations from Sol 135's final location. The miniature thermal emission spectrometer performed atmospheric measurements and an overnight observation during the early morning pass by NASA's *Mars Odyssey* orbiter. Part one of a planned ingress (entry) survey campaign with the panoramic camera was initiated.

MER-A Sol 157: Part of the Sol 157 plan was to observe End-of-Rainbow and use the alpha particle X-ray spectrometer, Mössbauer spectrometer and microscopic imager to study the Shredded soil target. However, the command load for Sol 157 never made it to *Spirit.* Further analysis indicated that the problem had to do with the frequency drift associated with the colder temperatures on Mars as the planet moves into its southern winter season. This was an anticipated problem, and the rover team has already imposed some strategies that will help to prevent the problem in the future.

—NASA/JPL

221 Makara 14 **2025 February 19 12:35 UTC**
Ls 046.4° Curiosity Sol 4458
MSD 53727 Perseverance Sol 1423

On this sol in 210 (2004 June 12):

MER-B Sol 137: *Opportunity* approached the rock target referred to as Tennessee. *Opportunity* drove 1.19 meters (3.9 feet) deeper into Endurance Crater, placing Tennessee perfectly within the instrument arm's reach. The rover is in position to perform the first series of arm operations starting on Sol 139 (210 Makara 16). Deep sleep mode was again invoked overnight from Sol 137 to Sol 138. The plans for the coming sols include grinding into Tennessee with the rock abrasion tool and investigating it with the rover's spectrometers.

MER-A Sol 158: *Spirit* got a break on Sol 157 and began Sol 158 with nicely charged batteries. She executed the activities originally planned for Sol 157, and then began to drive a bit closer to End-of-Rainbow. Although the direct path to End-of-Rainbow would be only a 4-meter (13 feet) drive, it was deemed too steep and hazardous by the rover team, so they planned a multi-stepped drive that could get the rover to the target safely. On Sol 158, the first part of the drive was completed, putting *Spirit* a little farther away from the End-of-Rainbow target, but with a straight shot to the rock for Sol 159 (210 Makara 15).

—NASA/JPL

221 Makara 15 **2025 February 20 13:14 UTC**
Ls 046.9° Curiosity Sol 4459
MSD 53728 Perseverance Sol 1424

On this sol in 145 (1882 March 13):

"Among the earlier drawings the longest radius of curvature, that is, the straightest representation of the Syrtis and the highest wind velocity, is that for 1882, March 13. The equivalent Martian date is given as May 2, and on the revised calendar it would be April 35. At this season the polar cap was melting very rapidly."

—William H. Pickering, 1918

On this sol in 162 (1914 March 06):

"On [Martian date] April 38 [according to the Pickering calendar], Cerberus was still brown and Propontis had vanished."

—William H. Pickering, 1928

On this sol in 210 (2004 June 13):

MER-B Sol 138: *Opportunity* spent Sol 138 grinding an 8.12-millimeter (0.32 inch) hole into the rock target called Tennessee. It took the rover 2 hours and 4 minutes to complete this grind, which is the deepest yet of the mission. After all this hard work, the rover went into a deep sleep for the night.

—NASA/JPL

221 Makara 16 **2025 February 21 13:54 UTC**
Ls 047.3° Curiosity Sol 4460
MSD 53729 Perseverance Sol 1425

Mars arrival opportunity from Earth via an EEM trajectory.

Earth arrival opportunity from Mars via a Type I trajectory.

On this sol in 210 (2004 June 14):

MER-B Sol 139: *Opportunity* used the microscopic imager to analyze the hole in Tennessee. The rover also performed both a daytime Mössbauer spectrometer observation and an overnight alpha particle X-ray spectrometer integration on the abraded surface.

—NASA/JPL

221 Makara 17 **2025 February 22 14:34 UTC**

Ls 047.8° Curiosity Sol 4461

MSD 53730 Perseverance Sol 1426

Possible meteor shower associated with parent body 5335 Damocles: Martian right ascension 140.4, declination 48.7.

On this sol in 153 (1897 April 01):

Part 7 of *The Martian* by George du Maurier appeared in *Harper's New Monthly Magazine* (date approximate).

On this sol in 174 (1936 October 01):

"The Human Pets of Mars" by Leslie F. Stone appeared in *Amazing Stories* (date approximate). Martians land on a golf course at Haines Point, Washington, D.C., and capture a small group of humans, transporting them back to Mars, where they are kept as pets.

Interior art by Harold Welch.

"Crystals of Madness" by D. L. James appeared in *Thrilling Wonder Stories* (date approximate). Humanity is menaced by an insidious, sentient life-force on the satellite Deimos.

Interior art by Marco Marchioni.

"The Tree of Life" by C.L. Moore appeared in *Weird Tales* (date approximate). A gripping tale of the planet Mars and the terrible monstrosity that called its victims to it from afar—a tale of Northwest Smith.

Interior art by Virgil Finlay.

On this sol in 186 (1959 April 27):

The song "The Little Martian" by the American songwriter, musician, and singer Jan Amber was released as a vinyl single.

On this sol in 210 (2004 June 15):

MER-B Sol 140: *Opportunity* was busy finishing up some miniature thermal emission spectrometer observations of Tennessee and then began a long Mössbauer integration that ran for the entire sol. *Opportunity* enjoyed some more deep sleep after these activities and will finish up the last of the Tennessee observations the morning of Sol 141.

MER-A Sol 161: *Spirit* used the microscopic imager to get an up-close view of Pot-of-Gold, but unfortunately, the images were out of focus. Rover planners attribute the blurry shots to a lower-than-expected rock contact prior to the imaging sequence. *Spirit* did successfully acquire Mössbauer observations of Pot-of-Gold on this sol.

—NASA/JPL

221 Makara 18 **2025 February 23 15:13 UTC**

Ls 048.2° Curiosity Sol 4462
MSD 53731 Perseverance Sol 1427

On this sol in 188 (1963 January 31):

Participating in NASA Marshall Space Flight Center's Early Manned Planetary – Interplanetary Roundtrip Expedition (EMPIRE) study, a General Dynamics study led by Krafft Ehricke issued its report describing a 15-month orbiter mission to be launched toward Mars in 1975 March.

On this sol in 205 (1995 January 20):

The song "The Martian Stomp" by the American surf rock band The Boss Martians was released on their eponymous CD.

On this sol in 210 (2004 June 16):

MER-B Sol 141: Opportunity completed post-rock abrasion tool Mössbauer spectrometer observations on the rock called Tennessee. Having spent the last four sols investigating Tennessee, Opportunity stowed its arm and moved deeper into Endurance Crater. A 0.70-meter (2.3 feet) drive positioned the vehicle to begin observations on the first contact point, a transition between two different geologic layers. As it turns out, after arriving at the rover's new location and taking images, there appears to be not one contact point, but three contact points all within reach of the arm. Opportunity then performed 2.5 hours of remote observations using the panoramic camera and miniature thermal emission spectrometer. The rover then went into deep sleep mode for the night.

MER-A Sol 162: *Spirit* was commanded to retake the microscopic images of Pot-of-Gold and do extensive observations with the Mössbauer and alpha particle X-ray spectrometer instruments. The microscopic images were successfully obtained, but a positioning fault of the instrument deployment device prevented the Mössbauer and alpha particle X-ray spectrometer observations from executing.

—NASA/JPL

221 Makara 19 **2025 February 24 15:53 UTC**

Ls 048.7° Curiosity Sol 4463
MSD 53732 Perseverance Sol 1428

On this sol in 210 (2004 June 17):

MER-B Sol 142: *Opportunity* began another series of microscopic imager observations on three different targets: Bluegrass, Siula Grande and Churchill. The rover then performed alpha particle X-ray spectrometer and Mössbauer spectrometer integrations overnight.

MER-A Sol 163: *Spirit* successfully completed the work with its instrument deployment and took some additional microscopic images of Pot-of-Gold.

Spirit currently sits at the base of the Columbia Hills in an area called the Hank Moore Hollow. This area has a collection of intriguing rocks on its rim, one of which, Pot-of-Gold, will be the first target for scientists' observations.

—NASA/JPL

221 Makara 20 **2025 February 25 16:32 UTC**

Ls 049.1° Curiosity Sol 4464
MSD 53733 Perseverance Sol 1429

On this sol in 210 (2004 June 18):

MER-B Sol 143: *Opportunity* again used its rock abrasion tool, but not without some consternation from the uplink team. In order to grind using the rock abrasion tool, a minimum of force must be used to push the tool onto the target. With the vehicle tilted 23.2 degrees, there was concern that applying too much force could cause the vehicle to lose traction on the slope and slide farther into the crater, possibly damaging the arm. Concerns were pacified when the rock abrasion tool operation worked flawlessly, abrading approximately 3 millimeters (about 0.12 inches) into the rock called Cobble Hill.

MER-A Sol 164: *Spirit* completed an overnight alpha particle X-ray spectrometer observation on Pot-of-Gold and then gathered additional microscopic images of the rock. In the afternoon, rover planners tried to reposition *Spirit* from a heading of about 170 degrees to a heading of about 95 degrees to improve the instrument deployment device positioning on Pot-of-Gold for the upcoming rock abrasion tool work. The planned traverse sent *Spirit* past Pot-of-Gold, down a slope on the west side of Hank's Hollow, then turned the rover to re-approach at the desired heading. The slope and slippage were greater than expected, leaving the rover at a westerly tilt of 19 degrees and almost 2 meters (6.6 feet) away from the target.

—NASA/JPL

221 Makara 21 **2025 February 26 17:12 UTC**

Ls 049.6° Curiosity Sol 4465
MSD 53734 Perseverance Sol 1430

On this sol in 210 (2004 June 19):

MER-B Sol 144: While *Opportunity* is hard at work inside Endurance Crater, engineers at JPL are busy testing engineering models in the Lab's simulated Martian environment. A tilt platform is being used to determine *Opportunity's* ability to climb back up over the "curb" below its current location.

Opportunity completed the Mössbauer spectrometer integration on the rock abrasion tool hole on the Cobble Hill area. Microscopic images were also collected.

MER-A Sol 165: *Spirit* worked toward getting into position in front of Pot-of-Gold, but given the slippage and need to avoid overshooting the target, engineers anticipated it would take at least two sols to get properly repositioned. Unfortunately, the drive made less progress than desired due in part to a collection of rocks encountered by the left rear wheel.

—NASA/JPL

221 Makara 22 **2025 February 27 17:52 UTC**
Ls 050.1° Curiosity Sol 4466
MSD 53735 Perseverance Sol 1431

On this sol in 172 (1933 January 01):

Part 2 of 3 of *The Wreck of the Asteroid* by Laurence Manning appeared in *Wonder Stories* (date approximate).

Interior art by Frank R. Paul.

On this sol in 184 (1955 July 28):

The *X Minus One* presentation of the radio drama "The Green Hills of Earth" by Robert A. Heinlein was broadcast. A song-writing space engineer is blinded in an accident; later, when nearing the end of his life, he wishes to make a final trip home to see the "green hills of earth." The broadcast included the poet Rhysling's "The Grand Canal" (see 220 Taurus 09).

On this sol in 210 (2004 June 20):

MER-B Sol 145: This was a busy sol, with *Opportunity* collecting more post-rock abrasion tool Cobble Hill and pre-Virginia microscopic images. The tool then bored a 4.3-millimeter (0.17 inches) hole in Virginia. Deep sleep mode was invoked for the overnight hours.

MER-A Sol 166: *Spirit* took advantage of its current position and used the gathered Mössbauer data and microscopic images of the soil in front of it before continuing the hill climb. *Spirit's* front wheels made it over the crest, leaving the rover at about a 13-degree tilt and still about 1 meter (3.3 feet) from Pot-of-Gold.

—NASA/JPL

221 Makara 23 **2025 February 28 18:31 UTC**
Ls 050.5° Curiosity Sol 4467
MSD 53736 Perseverance Sol 1432

On this sol in 210 (2004 June 21):

MER-B Sol 146: *Opportunity* examined the newly-abraded hole in Virginia with the microscopic imager and the alpha particle X-ray spectrometer.

MER-A Sol 167: *Spirit* looked at a bit of soil called Jaws with the alpha particle X-ray spectrometer and microscopic imager. Then the rover completed a drive intended to put it into position to analyze Pot-of-Gold with the instruments on its robotic arm. The drive moved *Spirit* farther than expected though, and the rover ended up directly over the rock. That position prevented any observations with the instrument deployment device.

—NASA/JPL

221 Makara 24 **2025 March 01 19:11 UTC**
Ls 051.0° Curiosity Sol 4468
MSD 53737 Perseverance Sol 1433

On this sol in 210 (2004 June 22):

MER-B Sol 147: *Opportunity* performed a long Mössbauer spectrometer integration on Virginia and completed some remote sensing from its location in the crater. After relaying the data through both *Mars Global Surveyor* and *Mars Odyssey, Opportunity* went into deep sleep mode for the night.

MER-A Sol 168: Rover planners commanded *Spirit* to "bump" backward, into a position where the rock abrasion tool could make contact with Pot-of-Gold. This was successful, and *Spirit* spent the rest of the time taking images of the surrounding area with its panoramic and navigation cameras.

—NASA/JPL

221 Makara 25 **2025 March 02 19:50 UTC**
Ls 051.4° Curiosity Sol 4469
MSD 53738 Perseverance Sol 1434

On this sol in 210 (2004 June 23):

MER-B Sol 148: London was *Opportunity's* target on Sol 148. The rock abrasion tool ground a 4.5-millimeter (0.18 inches) hole in the rock. The alpha particle X-ray spectrometer was then placed on the hole for integration.

MER-A Sol 169: *Spirit* successfully operated its rock abrasion tool on Pot-of-Gold, grinding away the top 0.2 millimeters (0.008 inches) of rock from the high points. The procedure took 1 hour and 45 minutes. Pot-of-Gold posed a special challenge to the rover team because it is quite small—only slightly larger than the rock abrasion tool instrument itself. The rock abrasion tool inflicts about 6.8 kilograms (15 pounds) of pressure on its rock targets, and smaller rocks aren't necessarily stable enough to resist this.

Before and after pictures of Pot-of-Gold showed that the rock was moved by the rock abrasion tool procedure. That movement, plus possible slippage where the tool contacts the rock, resulted in only intermittent contact during the grinding operation. After the grind was complete, *Spirit* placed the alpha particle X-ray spectrometer on the freshly exposed area in preparation for an operation later that night.

—NASA/JPL

221 Makara 26 **2025 March 03 20:30 UTC**

Ls 051.9° Curiosity Sol 4470
MSD 53739 Perseverance Sol 1435

On this sol in 154 (1899 February 27):

Konstantin Mikhailovich Bartashevich (Константин Михайлович Барташевич) was born; Soviet actor and assistant director; starred in the motion picture *Небо зовет (The Sky Calls)* (1959).

On this sol in 197 (1980 January 13):

Two 180-million-year-old olivine-phyric shergottite meteorites with a total mass of 7,942 grams designated EETA 79001 A & B were discovered in the Elephant Moraine, Antarctica. They were later identified as having originated on Mars.

On this sol in 210 (2004 June 24):

MER-B Sol 149: *Opportunity* continued to scrutinize London with the alpha particle X-ray spectrometer and the microscopic imager. A Mössbauer spectrometer integration was initiated and will be completed on Sol 150 (210 Makara 27). Deep sleep mode was invoked for the overnight hours.

MER-A Sol 170: *Spirit* awoke to stop the alpha particle X-ray integration, took miniature thermal emission spectrometer and panoramic camera images of some local track marks, took more microscopic images of the newly exposed Pot-of-Gold, then placed the Mössbauer instrument on the site for a 21-hour overnight observation.

—NASA/JPL

221 Makara 27 **2025 March 04 21:10 UTC**

Ls 052.3° Curiosity Sol 4471
MSD 53740 Perseverance Sol 1436

On this sol in 210 (2004 June 25):

MER-B Sol 150: *Opportunity* completed Mössbauer spectrometer observations in the rock abrasion tool hole on the rock Tennessee (hole number 3). The rover then stowed its arm, drove 0.55 meters (1.8 feet) backwards, turned to 40 degrees and then drove 1 meter (3.3 feet) forward. That was a net forward motion of 0.45 meters (1.5 feet) down-crater. This drive enabled *Opportunity* to reach two targets in the fifth distinct layer (E) of Endurance Crater. The night of Sol 150 into morning of Sol 151, Opportunity did not do a deep sleep.

MER-A Sol 171: *Spirit* continued its investigation in Hank's Hollow and the rock target Pot-of-Gold. The rover successfully completed observations of the abraded area with the microscopic imager and alpha particle X-ray spectrometer.

—NASA/JPL

221 Makara 28 **2025 March 05 21:49 UTC**

Ls 052.8° Curiosity Sol 4472
MSD 53741 Perseverance Sol 1437

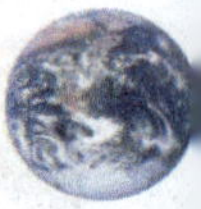

Approximately the last sol of the month Zeus on the Heinlein calendar; when Jim Marlowe found the Martian bouncer Willis in *Red Planet* (1949).

On this sol in 210 (2004 June 26):

MER-B Sol 151: *Opportunity* began the sol by performing panoramic camera images, then it unstowed its arm and used the microscopic imager to capture the next drilling target, Grindstone. After using the microscopic imager, *Opportunity* spent two hours grinding and created another precise hole. Upon completing the grinding operation, *Opportunity* placed the Mössbauer in the hole and performed a long integration. *Opportunity* did a mini-deep sleep operation overnight from Sol 151 into the morning of Sol 152. The miniature thermal emission spectrometer reached a chilly -51 degrees Celsius overnight.

MER-A Sol 172: *Spirit* looked at the sky with the miniature thermal emission spectrometer and panoramic camera. The rover finished up its Pot-of-Gold observations with some microscopic images and a final long Mössbauer integration of the abraded surface.

—NASA/JPL

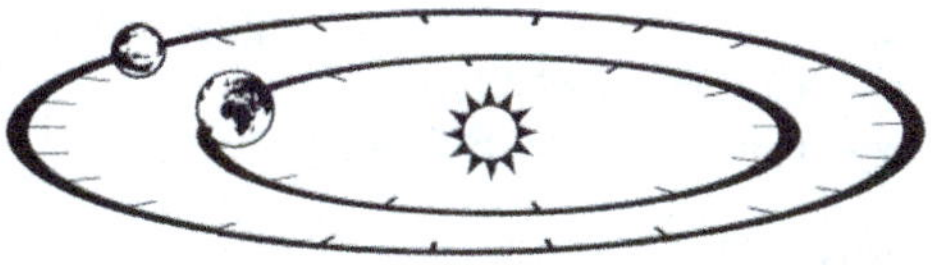

221 AQUARIUS

As seen from Mars, the Sun traverses the western sector of the constellation Aquarius, which in the Babylonian star catalogues was represented as the god Ea holding a vase overflowing with water.

The median one-way communications delay during this month is 8 minutes 38 seconds.

221 Aquarius 01 **2025 March 06 22:29 UTC**

Ls 053.2° Curiosity Sol 4473
MSD 53742 Perseverance Sol 1438

Approximately the first sol of November on the Heinlein calendar.

On this sol in 147 (1885 December 30):

Edwin Jerome Rath (Edwin Jerome) was born; American actor; starred in the *X Minus One* episode "The Martian Death March" (1955).

On this sol in 149 (1889 October 03):

William Elliott Dold, Jr. was born; American pulp science fiction artist; created the interior art for "Spoor of the Bat" by Arthur Leo Zagat (1934), "Old Faithful" by Raymond Z. Gallun (1934), "Son of Old Faithful" by Raymond Z. Gallun (1935), "Child of the Stars" by Raymond Z. Gallun (1936), and "The Secret of the Canali" by Clifton B. Kruse (1938).

On this sol in 210 (2004 June 27):

MER-B Sol 152: *Opportunity* completed the observations on the hole on Grindstone. When the rover woke up from deep sleep at 7:00 a.m. local solar time, it turned on the Mössbauer spectrometer and integrated until mid-afternoon. When the Mössbauer integration was complete, the rover switched tools to place the alpha particle X-ray spectrometer in the hole.

MER-A Sol 173: *Spirit* performed atmospheric observations with the miniature thermal emission spectrometer and panoramic camera. The rover also took some panoramic camera context images for the Sol 172 (210 Makara 28) thermal inertia observations. The rover finished the day's work by stowing the instrument deployment device and doing a "bump-back" to the Bread Box target. One last panoramic camera shot of Pot-of-Gold ended up a bit overexposed and will need to be retaken.

—NASA/JPL

221 Aquarius 02 **2025 March 07 23:08 UTC**

Ls 053.7° Curiosity Sol 4474
MSD 53743 Perseverance Sol 1439

On this sol in 210 (2004 June 27):

MER-B Sol 152: *Opportunity* acquired some thermal inertia observations of nearby soil with the miniature thermal emission spectrometer.

On this sol in 210 (2004 June 29):

MER-A Sol 174: *Spirit* began the sol acquiring atmospheric observations with the mini thermal emission spectrometer and panoramic camera. The rover then imaged the drive direction with the panoramic camera. Last but not least, *Spirit* took a look at a disturbed area of soil called Bright Tracks with the panoramic camera to help scientists learn more about the very bright material found here.

During the next 15 or more sols, rover planners will perform a 3,000-meter tune-up on *Spirit* before the rover embarks on a climb up the hills. The tune-up will include a number of elements including:

- A front hazard avoidance camera calibration where a series of robotic arm poses and hazard avoidance camera images will be used to refine the rover planners' ability to target objects using stereo hazard avoidance camera images. The team is currently experiencing a 2- to 3-centimeter (slightly less or slightly greater than an inch) error in predicted versus actual target locations in the vicinity of the instrument deployment device.

- *Spirit's* first deep sleep. Deep sleep is a mode that leaves the rover completely un-powered overnight, saving the energy that would be spent powering rover electronics and survival heaters that are normally on even when the rover is napping. *Spirit* needs deep sleep to save energy in the coming sols. Since deep sleep is potentially harmful to the mini thermal emission spectrometer instrument because its survival heater is not powered, rover planners have identified two observations that must be completed before the first deep sleep is attempted. *Opportunity* has been using deep sleep for several weeks now.

- A right front wheel lubrication. *Spirit's* right front wheel continues to draw roughly twice the current of the other wheels. *Spirit* will drive to Engineering Flats, a relatively flat, hazard-free area where rover planners will execute a series of diagnostic drive tests and heating sequences over the course of four to five sols. The intent is that the heating will re-flow the lubricants in this actuator, correcting the problem. Engineering Flats is roughly 7 meters (nearly 23 feet) from *Spirit's* current location.

- Engineering tests of visual odometry. Visual odometry uses navigation camera images taken during a drive to determine the rover's location. This rover feature has been improved and is ready for trial runs now. Rover planners would like to use it on a regular basis to get *Spirit* where they want it to go more quickly. Due to slippage, *Spirit* sometimes needs two or more sols to make a short approach when using the blind drive technique.

—NASA/JPL

221 Aquarius 03 **2025 March 08 23:48 UTC**

Ls 054.1° Curiosity Sol 4475
MSD 53744 Perseverance Sol 1440

On this sol in 149 (1889 October 05):

Robert Gibson Jones was born; American pulp artist; created the cover art for the 1950 September issue of *Amazing Stories* featuring "You Can't Escape from Mars!" by E.K. Jarvis.

On this sol in 210 (2004 June 29):

MER-B Sol 154: The sol consisted of *Opportunity* completing activities on the target Kettlestone, including a long Mössbauer integration, some microscopic images and placement of the arm for a little early morning alpha particle X-ray spectrometer integration to occur on the morning of Sol 155 (210 Aquarius 04). The rover then went to sleep.

On this sol in 210 (2004 June 30):

MER-A Sol 175: *Spirit* analyzed the new targets Breadbox and Sourdough with its panoramic camera and miniature thermal emission spectrometer. *Spirit* then got an up-close look at Breadbox with the microscopic imager, and deployed the Mössbauer spectrometer on Sourdough for an overnight integration. In the middle of the Martian night, *Spirit* did a tool change to the alpha particle X-ray spectrometer and completed a five-hour integration before the Sol 176 (210 Aquarius 04) plan began.

—NASA/JPL

221 Aquarius 04 **2025 March 10 00:27 UTC**

Ls 054.6° Curiosity Sol 4476
MSD 53745 Perseverance Sol 1441

As seen from Mars, the Sun enters the constellation Aquarius.

On this sol in 140 (1874 September 21):

Gustav Theodore Holst was born; British composer, arranger and teacher; composed the seven-movement orchestral suite *The Planets* (1918), of which "Mars, the Bringer of War" is the first movement.

On this sol in 210 (2004 June 30):

MER-B Sol 155: *Opportunity's* Sol 155 began with an early morning *Mars Odyssey* UHF relay of about 60 megabits of data, followed by a completion of the alpha particle X-ray spectrometer integration on Kettlestone. The rover then performed a calibration activity with the arm, consisting of moving the arm into about 20 different poses and imaging each pose with the front hazard-avoidance cameras. From the stereo images and the reported position of the arm, the rover team will be able to update models and better target the instruments onto surface features in the future. Some miniature thermal emission spectrometer activity was conducted midday, and then the rover drove backwards about 1 meter (3.3 feet). The drive backwards served two purposes: first, it positioned the rover to image the most recent rock abrasion tool holes with the panoramic camera and miniature thermal emission spectrometer; secondly, it gave the team an opportunity to evaluate driving back up over the "curb" that was considered so difficult before traversing on Sol 150 (210 Makara 27). The drive back up over the curb went very well. Slip was estimated at around 11 percent, admirable for such a traverse.

On this sol in 210 (2004 July 01):

MER-A Sol 176: *Spirit* spent the sol getting a battery re-charge and a front hazard avoidance camera calibration. That evening, engineers commanded *Spirit* to wake up and enable the panoramic camera mast actuator heater so they could determine when the thermostat turns the heater on. The heater turned on when expected, which will allow *Spirit* to conduct a night-time miniature thermal emission spectrometer observation in a few sols.

—NASA/JPL

221 Aquarius 05 — **2025 March 11 01:07 UTC**

Ls 055.0° — Curiosity Sol 4477

MSD 53746 — Perseverance Sol 1442

As seen from Mars, the Sun enters the constellation Aquarius.

On this sol in 152 (1895 June 01):

Part 2 of *Mars* by Percival Lowell appeared in *The Atlantic* (date approximate).

On this sol in 210 (2004 July 01):

MER-B Sol 156: Due to an incorrect time conversion, the rover team failed to get the intended command load to *Opportunity* at the right time. As a result, the spacecraft executed a backup set of minimal activities and returned about 80 megabits of data through *Mars Odyssey* in the afternoon.

On this sol in 210 (2004 July 02):

MER-A Sol 177: *Spirit* successfully performed a series of observations on an interesting and shiny feature called String of Pearls. The rover acquired two microscopic images of the target and an overnight integration with the Mössbauer spectrometer and alpha particle X-ray spectrometer. As Mars's southern winter approaches, *Spirit's* energy resources become increasingly limited. Overnight tool changes and their associated heating take a big toll on the limited energy budget and require some preparation and recovery to keep up *Spirit's* battery charge.

—NASA/JPL

On this sol in 173 (1934 December 01):

Part 2 of *Swords of Mars* by Edgar Rice Burroughs appeared in *The Blue Book Magazine* (date approximate).

Interior art by Robert Fink.

Part 1 of *The Titan* by P. Schuyler Miller appeared in *Marvel Tales of Science and Fantasy* (date approximate). Considers an advanced but decadent civilization on Mars, told from the viewpoint of the Martians. Includes passages about sexual relations between female aristocracy and male worker-slaves.

"Old Faithful" by Raymond Z. Gallun appeared in *Astounding Stories* (date approximate). Using light flashes, Martian 774 and Professor Walters have exchanged signals, laboriously extracting limited meaning from a minimum of light patterns. It is possible that in another ten years or so fluent, more complex communication might take place, but 774 has received a death sentence. On Mars, where resources are very limited and strictly rationed, individuals live an allotted time, with extensions granted only for work useful to the culture.

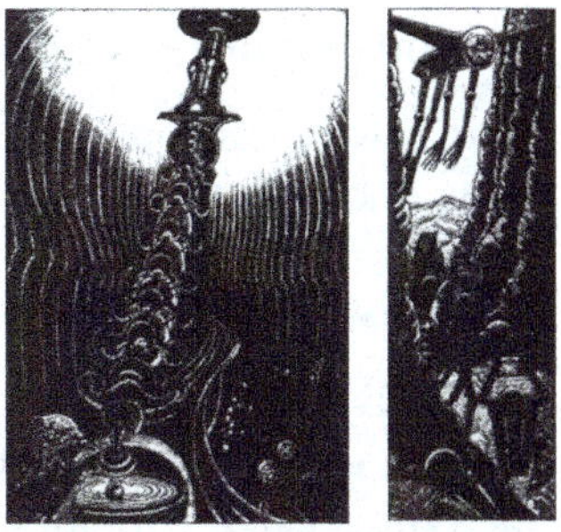

—Bleiler and Bleiler, 1998

Interior art by Elliott Dold, Jr.

On this sol in 210 (2004 July 02):

MER-B Sol 157: *Opportunity* acquired some images of the rock abrasion tool holes from previous sols. Then it drove down the hill to approach the next target. It drove beautifully and achieved its goal location. However, due to the large slopes (final rover tilt was 28.6 degrees), *Opportunity* ended the drive with the right rear wheel apparently slightly above the terrain (not touching anything). Even in this state the rover appears to be stable, but the team will likely take action on the next sol to get the suspension squared up (six wheels touching) before proceeding with preparations to grind with the rock abrasion tool again.

On the night of Sol 157 to 158 the rover gave up deep sleep in order to preserve an exceptional morning *Mars Odyssey* pass.

On this sol in 210 (2004 July 03):

MER-A Sol 178: *Spirit* began the sol by stowing the robotic arm and then backing up 1.5 meters (about 5 feet) from Hank's Hollow in order to properly place the miniature thermal emission spectrometer to get a good view of Pot of Gold and nearby rover tracks. Engineers also took this opportunity to use visual odometry for the first time on *Spirit.* This is a technique in which the rover takes successive images of its surroundings during a drive and then matches features in those images on-board to compute how far and in what direction it has moved. Both the drive and the test went well, and ground verification showed that the matching worked quite nicely with the features in this terrain. Visual odometry will be important if and when *Spirit* starts driving on five wheels, since the actual drives can and will be rather different than what is commanded. The rover can use the visual odometry estimates while driving to compensate for the slipping and yawing which engineers expect with five-wheel driving.

—NASA/JPL

221 Aquarius 07 **2025 March 13 02:26 UTC**

Ls 055.9° Curiosity Sol 4479

MSD 53748 Perseverance Sol 1444

On this sol in 148 (1887 November 23):

William Henry Pratt (Boris Karloff) was born; British actor; starred in the *Tales of Tomorrow* episode "Memento" (1952).

On this sol in 176 (1940 July 24):

Edgar Rice Burroughs began writing "John Carter and the Pits of Horz," first of a series for new book. The story was later was published under the title "The City of Mummies."

On this sol in 210 (2004 July 03):

MER-B Sol 158: In the very early morning, *Opportunity* woke up to chat with the *Mars Odyssey* spacecraft and returned over 100 megabits of data. The rover then started the sol's activities early with an attempt to image clouds around 8:30 in the morning. It then went back to sleep until about 10:30. After the morning uplink, it acquired some microscopic images of the new target area, then stowed its arm to allow a small mobility maneuver to get all six wheels squarely planted on the ground. This seemed to go as planned and reduced the total tilt of the vehicle to only 26.4 degrees, but did not appreciably change its position. This left the rover, as desired, in position to perform science investigations on the next targets of interest.

On this sol in 210 (2004 July 04):

MER-A Sol 179: *Spirit* woke up at a new location and spent the day performing remote sensing with the miniature thermal emission spectrometer, including an overnight observation.

—NASA/JPL

221 Aquarius 08 **2025 March 14 03:06 UTC**

Ls 056.4° Curiosity Sol 4480
MSD 53749 Perseverance Sol 1445

On this sol in 210 (2004 July 04):

MER-B Sol 159: The operations team planned this sol's activity a day in advance so that they could a enjoy a much-deserved holiday on July 4th, resulting in a relatively quiet sol on Mars for *Opportunity*. The sol's activities included daily miniature thermal emission spectrometer and panoramic camera atmospheric observations. A calibration of the miniature thermal emission spectrometer vertical scan mirror actuator was also conducted. The vertical scan mirror actuator operates like a periscope, allowing the miniature thermal emission spectrometer to target things vertically from the ground to the sky.

On this sol in 210 (2004 July 05):

MER-A Sol 180: The rover has survived two times the original planned mission duration of 90 sols. On this notable sol, the rover continued with remote sensing, performing miniature thermal emission spectrometer observations on disturbed soil and rover tracks. *Spirit* then looked at the targets Cookie Cutter and Julienned with the panoramic camera. Because of power and timing issues, *Spirit* was not able to complete intended microscopic imaging, Mössbauer spectrometer, and alpha particle X-ray spectrometer measurements at this site. These operations were moved into the Sol 181 (210 Aquarius 09) plan. Total odometry after Sol 180 is 3,414 meters (2.1 miles).

—NASA/JPL

221 Aquarius 09 **2025 March 15 03:45 UTC**

Ls 056.9° Curiosity Sol 4481
MSD 53750 Perseverance Sol 1446

The first sol of June, marsrev 3679 on the Levitt calendar.

On this sol in 171 (1931 March 01):

"On Board the Martian Liner" by Miles J. Breuer, M.D. appeared in *Amazing Stories* (date approximate). When space travel is an accepted matter—who knows but that it might be soon—there must undoubtedly be much room for adventure—not only what we know would be incident to the danger of moving through space, or what would be due to the unknown factors that might be met with on landing. Because a man is going through space, it is not a foregone conclusion that he will become all that is noble. Neither will mysterious offenses be confined to the earth. But we can't send a detective on every ship.

Interior art by Leo Morey.

On this sol in 210 (2004 July 05):

MER-B Sol 160: *Opportunity* forfeited deep sleep overnight from Sol 159 to Sol 160 to take advantage of an optimal communication window with *Mars Odyssey.* We used the microscopic imager to take a mosaic of the Drammensfjorden location on the rock Millstone, which is in layer F of Endurance Crater. The microscopic images were taken to enable the accurate placement of the rock abrasion tool on Sol 161. The Mössbauer spectrometer was then placed on the compositional calibration target (CCT). This was the first such use of the CCT and was done partially out of concern that the instrument's behavior might be affected by the rover's present tilt of roughly 25 degrees. The rover team put *Opportunity* into a deep sleep the night of Sol 160.

On this sol in 210 (2004 July 06):

MER-A Sol 181: The plan for *Spirit* was to deploy the instrument deployment device for microscopic imaging, then perform a two-hour Mössbauer integration. The rover was to conduct miniature thermal emission spectrometer and navigation camera observations of rover-disturbed soil. After this, the rover was to drive to a relatively flat area dubbed Engineering Flats. This was to prepare for a multi-sol engineering activity to heat, and ideally re-lubricate, the right-front wheel actuator. After the drive, *Spirit* was to take a 360-degree navigation camera panorama, followed by miniature thermal emission spectrometer observations during the communications session with NASA's *Mars Odyssey* orbiter. Unfortunately, as *Spirit* began to execute the Sol 181 plan, the onboard software predicted an instrument deployment device collision. This prevented further arm functions and the drive.

—NASA/JPL

---

221 Aquarius 10 **2025 March 16 04:25 UTC**

Ls 057.3° Curiosity Sol 4482
MSD 53751 Perseverance Sol 1447

On this sol in 210 (2004 July 06):

MER-B Sol 161: *Opportunity* awoke from deep sleep to make some early morning atmospheric observations, including another attempt to image clouds. Later that morning the Mössbauer instrument was stopped and removed from the CCT. The rock abrasion tool was then used on the target Drammensfjorden, creating a 6.3mm (a quarter of an inch)-deep hole during the two-and-one-half hour sequence. The day ended with more atmospheric observations and a placement of the alpha particle X-ray spectrometer on the abraded rock abrasion tool hole.

On this sol in 210 (2004 July 07):

MER-A Sol 182: Rover planners quickly determined the cause of the instrument deployment device error on *Spirit* and continued to plan for Sol 182 as normal. The intent for Sol 182 was to complete the activities originally intended for Sol 181. The Sol 182 plan executed nominally, placing *Spirit* squarely in the middle of Engineering Flats.

—NASA/JPL

221 Aquarius 11 **2025 March 17 05:05 UTC**
Ls 057.8° Curiosity Sol 4483
MSD 53752 Perseverance Sol 1448

On this sol in 210 (2004 July 07):

MER-B Sol 162: *Opportunity* woke up in the wee hours of the sol for a *Mars Odyssey* communication session and to start the alpha particle X-ray spectrometer observation. The alpha particle X-ray spectrometer collected data from the hole in Drammensfjorden until 10:30 a.m. local solar time. The data indicated that the alpha particle X-ray spectrometer doors were fully open, despite the failure of the door-open switch to trigger when the doors were opened on the CCT. The rover then took a series of microscopic images of the rock abrasion tool hole before starting a Mössbauer integration at the same location. The Mössbauer integrated until the team invoked deep sleep at 7 p.m. local solar time, and was restarted after deep sleep at 7 a.m. the next sol.

—NASA/JPL

221 Aquarius 12 **2025 March 18 05:44 UTC**
Ls 058.2° Curiosity Sol 4484
MSD 53753 Perseverance Sol 1449

On this sol in 210 (2004 July 08):

MER-B Sol 163: Rover engineers stopped *Opportunity's* Mössbauer, then successfully opened and closed the alpha particle X-ray spectrometer doors for diagnostic purposes. The rover arm was then stowed and the rover began a two-sol panoramic camera science survey of areas upslope from our current position.

—NASA/JPL

221 Aquarius 13 **2025 March 19 06:24 UTC**
Ls 058.7° Curiosity Sol 4485
MSD 53754 Perseverance Sol 1450

On this sol in 210 (2004 July 10):

MER-A Sol 185: On Sols 184 and 185, *Spirit* cycled heaters and attempted to lubricate the right front wheel's drive motor. *Spirit* also performed very short test drives. *Spirit's* orientation on Engineering Flats had the rover in a slight southerly tilt, away from the Sun. This low Sun angle, coupled with the power required to energize the heaters, put quite a strain on the batteries. The state of charge of the batteries after the test was very low, so science observations were not performed during these two sols.

—NASA/JPL

221 Aquarius 14 **2025 March 20 07:03 UTC**
Ls 059.1° Curiosity Sol 4486
MSD 53755 Perseverance Sol 1451

On this sol in 210 (2004 July 11):

MER-A Sol 186: *Spirit* used its instrument deployment device, or robotic arm, to validate a new front hazard-avoidance camera model. This new model will improve the positioning accuracy of the tools on the arm. The arm attempted to place the Mössbauer spectrometer at nine different locations. Seven targets were hit, and two targets were missed. The engineering team planned to repeat this operation on the two missed positions on a later sol. *Spirit* also performed a test drive to characterize the results of the lubrication activity. The beginning orientation of the vehicle had the right front wheel facing the Sun, so the starting temperature was much warmer than the original baseline test drive. This temperature difference makes it hard to accurately compare the pre- and post-test-drive results. Analysis indicates there is approximately a 20 percent increase in wheel drive performance, but engineers cannot attribute this gain to the four-sol wheel heating operation alone. *Spirit* sat mostly motionless for the about the last 30 sols, and that allowed some lubrication to re-flow naturally. The bottom line is that the right-front wheel's performance has improved, but it is still drawing about twice as much current as any other wheel.

—NASA/JPL

221 Aquarius 15 **2025 March 21 07:43 UTC**

Ls 059.6° Curiosity Sol 4487

MSD 53756 Perseverance Sol 1452

The first sol of the fourth quarter of Spring on the Aitken calendar.

On this sol in 176 (1940 August 01):

"Lost Treasure of Mars" by Edmond Hamilton also appeared in *Amazing Stories* (date approximate). Gareth Crane faced death because of one treasure cache. Was it a good idea to gamble his life of the chance of finding a greater one?

Interior art by Julian S. Krupa.

"Vault of the Beast" by A.E. van Vogt appeared in *Astounding Science Fiction* (date approximate). Beings from another dimension have sent a living plastic "robot" to Earth to find the "greatest mathematical mind in the Solar System," and get that person to open a vault on Mars, containing one of the race of its creators. It is able to imitate any form of matter, and to tap the thoughts of the being it duplicates. The creature kills its way to one man, Jim Brender, who it believes is the man. The creature, in the form of another man, reveals that the Martian vault was built by the Ancient Martians, made up of an "ultimate metal." The vault is known as the "Tower of the Beast," located in a buried Martian city. It says that the key to opening it is "factoring the ultimate prime number."

—Wikipedia

Time is running out. The time-lock on the Beast's millennia-old Martian cage is about to open, and mankind will never be able to deal with the threat!

—Goodreads.com

Illustration by Edd Cartier.

"There Was No Paradise" by Manley Wade Wellman appeared in *Thrilling Wonder Stories* (date approximate). The man from Mars had the Midas touch, but gold couldn't buy the things he wanted!

Interior art uncredited.

On this sol in 198 (1981 December 17):

The American punk rock band The Misfits performed their song "I Turned into a Martian" at The Ritz in New York City.

Oh go
Possession of the mind is a terrible thing
It's a transformation with an urge to kill
Not the body of a man from earth
Not the face of the one you love, 'cause

Well, I turned into a Martian
Whoa oh oh
I can't even recall my name
Whoa oh oh
Times I never hardly sleep at night
Whoa oh oh

Well, I turned into a Martian today
I walk down city streets
On an unsuspecting human world
Inhuman in your midst
This world is mine to own, 'cause

Go go
Well, I turned into a Martian
Whoa oh oh
Well, I can't even recall my name
Whoa oh oh
Times I never hardly sleep at night
Whoa oh oh
Well, I turned into a Martian today

Well, I turned into a Martian
Whoa oh oh
Well, I can't even recall my name
Whoa oh oh
Times I never hardly sleep at night
Whoa oh oh
Turned into a Martian
Whoa oh oh
Can't even recall my name
Oh, won't you tell me what the fuck is my name, Martian
Whoa oh oh
Whoa oh oh

Songwriter: Glenn Danzig

On this sol in 210 (2004 July 11):

MER-B Sol 166: The sol's tasks for *Opportunity* included imaging of possible traverse paths inside Endurance Crater, then the start of a long period of data collection by the Mössbauer spectrometer on a target called Dahlia. All went as planned.

On this sol in 210 (2004 July 12):

MER-A Sol 187: After the tune-up, *Spirit* was free to begin its drive away from Engineering Flats and head to higher ground and a better solar orientation. With its batteries very low, *Spirit* limped 8 meters (26 feet) to a location with a slightly better tilt toward the Sun and performed about 50 minutes of science observations.

—NASA/JPL

221 Aquarius 16 **2025 March 22 08:23 UTC**

Ls 060.0° Curiosity Sol 4488
MSD 53757 Perseverance Sol 1453

On this sol in 183 (1953 October 02):

The animated short film "Popeye, the Ace of Space" was released.

A semi-remake of the earlier Popeye cartoon "Rocket to Mars," Popeye drives along a country road in his boat-shaped car. When he sees that a bridge is out, he takes care of the problem with the help of some spinach. Meanwhile, an alien spaceship approaches Earth with its occupants bent on finding and abducting an average Earthman to conduct their tests on. They find Popeye instead, and snatch him away from his car via a propeller-mounted plunger. Later, the ship arrives at its planet of origin (determined by Popeye to be Mars).

Many green-skinned aliens are in attendance as the Earthman is put on the first of their sinister machines: a Cosmic Ager that soon makes him turn 125. Not amused by the experiment, he counters by eating his wonder vegetable, and reverts to age two. After the child regurgitates some excess spinach, he goes back to age 40. Next, he is strapped to a cruel contraption that would drop the huge "Atom Apple Smasher" upon his outstretched neck. Perusing a third can of spinach makes his Adam's apple as hard as Gibraltar and the massive hammer is split in two. The abductors then successfully try disintegrating the powerful being and, when his last spinach can is all that is left of him, the aliens fight for the source of his great might. Popeye is not entirely gone, however, and his invisible hand secures the spinach. Visible and strong again, he joins the spacemen's fight until they are finally pacified. He then borrows their spaceship and safely travels back to Earth while singing his song.

—Wikipedia

On this sol in 186 (1959 May 24):

Alfred Hitchcock Presents, Season 4 Episode 32, "Human Interest Story" airdate. Teleplay by Fredric Brown based on his short story "The Last Martian." Newspaperman Bill Everett is told by his editor to go to the bar across the street and interview a man who claims to be a Martian. There, Everett meets Howard Wilcox who spins a long tale about how he woke up one morning to find that his fellow Martians had all disappeared. He traveled to Earth and found himself in Wilcox's body. Everett convinces him to go home to his wife and even offers to accompany him. He nearly convinces Wilcox to keep his story to himself, but when he decides to tell his wife the whole story, Everett must take drastic action. All is explained when Everett provides a complete report to his editor.

—International Movie Database

On this sol in 210 (2004 July 12):

MER-B Sol 167: The sol saw completion of *Opportunity's* Mössbauer spectrometer's long integration at Dahlia, acquisition of some microscopic imager pictures, and placement of the alpha particle X-ray spectrometer on the capture magnet, which is one of the two magnets on the front of the rover deck. In the early morning hours of Sol 168, the alpha particle X-ray spectrometer collected data at this magnet. The composition of material sticking to the magnet is what interests scientists.

On this sol in 210 (2004 July 13):

MER-A Sol 188: *Spirit* continued to drive away from Engineering Flats. As a strategy for dealing with the right-front wheel, *Spirit* is now driving backwards and dragging its right front wheel when it is on relatively flat terrain. This strategy aims to extend the lifetime of the wheel's drive motor for use when it is needed most. *Spirit* performed its first backward test drive of 4 meters (13 feet) on this sol. *Spirit* also performed about an hour and a half of remote-sensing observations using the panoramic camera, navigation camera and miniature thermal emission spectrometer.

—NASA/JPL

221 Aquarius 17 **2025 March 23 09:02 UTC**

Ls 060.5° Curiosity Sol 4489
MSD 53758 Perseverance Sol 1454

On this sol in 210 (2004 July 13):

MER-B Sol 168: *Opportunity* lifted the alpha particle X-ray spectrometer off the capture magnet and replaced it with the Mössbauer spectrometer for the start of a long integration with that instrument. These complementary measurements of the material on the capture magnet should provide insight into the composition and magnetic properties of the dust around the rover. Remote-sensing activities were also performed.

Up to this point, *Opportunity* had not been commanded to take any images during an ultra-high-frequency (UHF) relay session with orbiting spacecraft. Testing prior to launch suggested that there could be electromagnetic interference that would degrade the telemetry link, so operations had carefully kept those activities separate. However, time could be used more efficiently if the rover could simultaneously take images and communicate. As an experiment, the rover was commanded to take navigation camera and panoramic camera images while transmitting on Sols 167 and 168, respectively. The quality of data sent during use of the navigation camera has been analyzed, and there seems to have been no ill effect. Data is still being analyzed from the transmission during use of the panoramic camera. The UHF relay session returned the expected amount of data. In both cases, no degradation of images was expected or seen. Based on this experiment, the engineering team will consider lifting the restriction against imaging during a UHF session.

On this sol in 210 (2004 July 14):

MER-A Sol 189: *Spirit's* battery state of charge increased due to a better tilt of the solar arrays toward the Sun. *Spirit* performed a precision 6-meter (20-foot), 6-wheel drive, then drove another 3 meters (10 feet) doing the wheel drag. *Spirit* also imaged an interesting rock outcrop. The outcrop was directly beneath the vehicle and extending northward.

—NASA/JPL

221 Aquarius 18 **2025 March 24 09:42 UTC**

Ls 060.9° Curiosity Sol 4490
MSD 53759 Perseverance Sol 1455

On this sol in 188 (1963 March 01):

Participating in NASA Marshall Space Flight Center's Early Manned Planetary - Interplanetary Roundtrip Expedition (EMPIRE) study, Lockheed Missiles and Space Division issued its report describing a 22-month flyby mission to be launched toward Mars on 1974 September 24.

The 1964 Mars Mariner B mission was replaced by an Atlas-Agena-launched, lightweight unmanned spacecraft called Mariner C. The mission was later renamed Mariner Mars 1964, which would be comprised of two spacecraft: Mariner C and Mariner D, ultimately launched as *Mariner 3* and *Mariner 4.*

On this sol in 210 (2004 July 14):

MER-B Sol 169: *Opportunity* drove deeper into the crater. A judgment had been made that the terrain in front of the rover would be no more difficult to traverse than terrain the rover had already crossed. Further, the science team was very interested in some geologic features about 3 to 4 meters (about 10 to 13 feet) down the slope, next to and including a rock called Knossos. The rover stowed its arm and trundled down the nearly 30-degree slope, arriving on a more-level area exactly where engineers intended. The rover is now below the steepest part of the inner slope in this part of the crater.

On this sol in 210 (2004 July 15):

MER-A Sol 190: *Spirit* completed remote sensing and a 13-meter (43-foot) drive, which included driving on 5 wheels to minimize further degradation of the actuator on the sixth, aging, right front wheel. During the 5-wheel portion of drive, visual odometry was used to accurately estimate the rover position. The drive ended with a short 6-wheel drive to achieve the desired position. This approach worked well, and engineers will continue to do 5-wheel driving to preserve the right front wheel actuator life while still achieving the desired position.

—NASA/JPL

221 Aquarius 19 **2025 March 25 10:21 UTC**

Ls 061.4° Curiosity Sol 4491
MSD 53760 Perseverance Sol 1456

Earth departure opportunity to Mars via an EVM trajectory.

Mars arrival opportunity from Earth via an EEM trajectory.

Earth arrival opportunity from Mars via a Type I trajectory.

On this sol in 148 (1889 October 22):

John L. Balderston was born; American screenwriter; co-wrote the screenplay for the motion picture *Red Planet Mars* (1952).

On this sol in 153 (1897 May 01):

Part 8 of *The Martian* by George du Maurier appeared in *Harper's New Monthly Magazine* (date approximate).

On this sol in 195 (1976 May 01):

The song "Martian Landscape" by the British rock band UFO was released on the vinyl LP *No Heavy Petting* (date approximate).

Come from the land
Where the wheat is still wheat
Where the water in the river is still water
In this land every man lends
A hand when you need it
But your hands are full of life
So you can help yourself and feel it

It's been a long way from the red of
my home
From the valleys
And the green plains never ending
Everyone knows that the sun in his
land is stronger
And your hands are full of life
If you can only wait just longer

So let it be a small haven
Big in your heart
Don't let it turn into a nova
Tryin' to climb and climb
Try and forget this place
Where I can still find time
To sing in rhyme

Martian landscape
I'll learn to lie in your arms
Where all the freedom is mine and it comes
Martian landscape
I'll turn my face to your sun
Where your horizon is mine and it comes

Come from the land
Where the wheat is still wheat
Where the water in the river is still water
In this land every man lends
A hand when you need it
But your hands are full of life
So you can help yourself and feel it
So let it be a small haven
Big in my heart
Don't let it turn into a nova
Tryin' to climb and climb
Try and forget this place
Where I can still find time
To sing in rhyme

Martian landscape
I'll learn to lie in your arms
Where all the freedom is mine and it comes
Martian landscape
I'll turn my face to your sun
Where your horizon is mine and it comes
Martian landscape
Martian landscape
Martian landscape
Martian landscape

Songwriters: Andy Parker / Phil Mogg / Danny Peyronnel

On this sol in 210 (2004 July 15):

MER-B Sol 170: Opportunity used panoramic camera to image Razorback and "Flatland" (a clean patch of bedrock nearby).

On this sol in 210 (2004 July 16):

MER-A Sol 191: *Spirit* took pictures with the microscopic imager and used the Mössbauer spectrometer on soil, then successfully completed another 10-meter (33-foot), 5-wheel drive to the north. (As *Spirit* is driving backwards to 'drag' the right front wheel, the robotic arm is effectively now in the back of the rover.)

—NASA/JPL

221 Aquarius 20 **2025 March 26 11:01 UTC**

Ls 061.8° Curiosity Sol 4492
MSD 53761 Perseverance Sol 1457

On this sol in 210 (2004 July 16):

MER-B Sol 171: *Opportunity* drove 3.7 meters (about 12 feet). Total odometry is now 1,478 meters (just over nine-tenths of a mile). Used the miniature thermal emission spectrometer to analyze some nearby geologic features. Took a 360-degree navigation camera mosaic.

On this sol in 210 (2004 July 17):

MER-A Sol 192: *Spirit* turned east off of a planned northerly traverse route to investigate a potential rock outcropping. The 17-meter (56-foot) drive, which included two sections of

5-wheel driving, landed *Spirit* right on top of the outcrop. Unfortunately, the drive left the rover at a bad tilt angle for solar energy, decreasing the available energy.

—NASA/JPL

221 Aquarius 21 **2025 March 27 11:40 UTC**
Ls 062.3° Curiosity Sol 4493
MSD 53762 Perseverance Sol 1458

On this sol in 210 (2004 July 17):

MER-B Sol 172: *Opportunity* used the panoramic camera and miniature thermal emission spectrometer instruments to image the solar panels and the miniature thermal emission spectrometer instrument's calibration target as part of a continuing evaluation of dust accumulation. Turned on the rover inertial measurement unit during the afternoon *Mars Odyssey* communication-relay pass as an experiment in support of our "teach your dog new tricks" campaign. If the inertial measurement unit does not adversely affect the communication, the team may be able to turn the rover during the communication relay sessions to increase the data return.

On this sol in 210 (2004 July 18):

MER-A Sol 193: *Spirit* completed a series of microscopic images and a short Mössbauer reading on the outcrop.

—NASA/JPL

221 Aquarius 22 **2025 March 28 12:20 UTC**
Ls 062.7° Curiosity Sol 4494
MSD 53763 Perseverance Sol 1459

On this sol in 188 (1963 March 05):

NASA issued "Request for Proposal No. 10-929, Voyager Design Studies" to 21 companies. A pair of unmanned orbiter/lander missions were to be launched on Saturn C-IB/Centaur launch vehicles to Venus in 1967 and to Mars in 1969.

On this sol in 210 (2004 July 19):

MER-B Sol 173: This view from the Mars Exploration Rover *Opportunity's* panoramic camera is an approximately true color rendering of the first seven holes that the rover's rock abrasion tool dug on the inner slope of Endurance Crater. The rover was about 12 meters (about 39 feet) down into the crater when it acquired the images combined into this mosaic. The view is looking back toward the rim of the crater, with the rover's tracks visible. The tailings around the holes drilled by the rock abrasion tool, or RAT, show evidence for fine-grained red hematite similar to what was observed months earlier in Eagle Crater outcrop holes.

Starting from the uppermost pictured (closest to the crater rim) to the lowest, the rock abrasion tool hole targets are called Tennessee, Cobblehill, Virginia, London, Grindstone, Kettlestone, and Drammensfjorden. *Opportunity* drilled these holes on Sols 138 (June 13, 2004), 143 (June 18), 145 (June 20), 148 (June 23), 151 (June 26), 153 (June 28) and 161 (July 7), respectively. Each hole is 4.5 centimeters (1.8 inches) in diameter.

This image was generated using the panoramic camera's 750-, 530-, and 430-nanometer filters.

MER-A Sol 194: *Spirit* took a large microscopic imager mosaic, consisting of 34 images at multiple positions, of a target called Sabre on an outcrop rock called Wooly Patch. This was followed by a two-hour reading by the Mössbauer spectrometer and an overnight, seven-hour reading by the alpha particle X-ray spectrometer.

—NASA/JPL

221 Aquarius 23 **2025 March 29 13:00 UTC**

Ls 063.2° Curiosity Sol 4495
MSD 53764 Perseverance Sol 1460

On this sol in 210 (2004 July 20):

MER-B Sol 174: *Opportunity* completed close-up examination of a rock target called Arnold Ziffel using the rover's microscopic imager, Mössbauer spectrometer and alpha particle X-ray spectrometer. All observations were successful.

MER-A Sol 195: *Spirit's* rock abrasion tool dug a surprisingly deep hole in only two hours of grinding. The rock appears to be softer than what scientists and engineers have seen previously in Gusev Crater. The alpha particle X-ray spectrometer was placed in the rock abrasion tool hole at Sabre. However, due to uncertainties in how long the arm and grinding operations take, the sequence was terminated a few minutes too early and a planned overnight alpha particle X-ray spectrometer integration did not take place.

—NASA/JPL

On this sol in 172 (1933 February 01):

"Escape from Phobos" by Neil R. Jones appeared in *Wonder Stories* (date approximate). Em Hanlek a murderer, is awaiting execution in the prison on Phobos when a fellow prisoner (Hoenbraun) contacts him and offers him a chance of escape As Hoenbraun gradually reveals, they will tunnel out of their cells to the storeroom, don space suits, strap on rockets and parachutes, and drop down to Mars. It is chancy, but still better than certain execution.

—Bleiler and Bleiler, 1998

Interior art by Frank R. Paul.

"At Bay in the Void" by D. D. Sharp also appeared in *Wonder Stories* (date approximate).

Interior art by Frank R. Paul.

Part 3 of 3 of *The Wreck of the Asteroid* by Laurence Manning also appeared in *Wonder Stories* (date approximate).

Interior art by Frank R. Paul.

On this sol in 175 (1938 September 23):

Out of the Silent Planet by C.S. Lewis was published. In the first novel of his classic science fiction trilogy, Dr Ransom, a Cambridge academic, is abducted and taken on a spaceship to the red planet of Malacandra, which he knows as Mars. His captors are plotting to plunder the planet's treasures and plan to offer Ransom as a sacrifice to the creatures who live there.

Ransom discovers he has come from the 'silent planet' – Earth – whose tragic story is known throughout the universe...

—Goodreads.com

Cover art by Harold Jones.

On this sol in 210 (2004 July 21):

MER-B Sol 175: Leaving Arnold Ziffel, *Opportunity* backed up to capture an image the results of the close-up work, and then moved on down slope to a new target. This move will leave the rover about 13 meters (about 43 feet) down from the lip of the crater. The rover was put into deep-sleep, energy-saving status overnight from Sol 175 to Sol 176.

MER-A Sol 196: *Spirit* recovered the alpha particle X-ray spectrometer integration. The sol began with a microscopic imager mosaic of the rock abrasion tool hole. The alpha particle X-ray spectrometer was put back in position in the hole, and reading lasting more than six hours was successfully performed, starting at about 4 a.m. Gusev time on Sol 197.

—NASA/JPL

---

221 Aquarius 25 **2025 March 31 14:19 UTC**
Ls 064.1° Curiosity Sol 4497
MSD 53766 Perseverance Sol 1462

On this sol in 210 (2004 July 22):

MER-B Sol 176: The sol was a driving day. *Opportunity* moved down the slope (and east) to a position to investigate a target called Diamond Jenness. Everything executed as planned, leaving the rover in a great position to grind into the target with its rock abrasion tool.

The engineering team is looking into a concern about the driving surface. The downslope pavement requires close examination before the rover traverses, to ensure the sand covering the pavement is still capable of supporting *Opportunity*. Recent experience has shown up to 30 percent slip.

MER-A Sol 197: *Spirit's* work on Sabre was completed with a very long, 21-hour Mössbauer integration, which was expected to be completed the morning of Sol 198. Before the integration was started, a microscopic imager mosaic was taken of Mammoth, the next rock abrasion tool target on Wooly Patch.

—NASA/JPL

---

221 Aquarius 26 **2025 April 01 14:58 UTC**
Ls 064.5° Curiosity Sol 4498
MSD 53767 Perseverance Sol 1463

Ante dies IV Idus Maius on the Roman calendar (analogous to May 25 on the recalibrated Pickering calendar), the anniversary of the dedication of the Templum Martis Ultoris (Temple of Mars the Avenger) in the Forum Augusti (Forum of Augustus) in Rome. The emperor Augustus built the temple in celebration of his victory over the two main conspirators in the assassination of Gaius Julius Caesar—Marcus Junius Brutus and Gaius Cassius Longinus—at the battle of Philippi.

On this sol in 141 (1874 October 03):

Charles Brown Middleton was born; American actor; starred in the serial motion picture *Flash Gordon's Trip to Mars* (1938).

On this sol in 210 (2004 July 23):

MER-B Sol 177: *Opportunity* performed a two-hour rock abrasion tool grind on the target Diamond Jenness, then took the resulting hole's picture with the microscopic imager. Surface debris and the bumpy shape of the rock apparently contributed to a shallow and irregular hole, only about 2 millimeters or 0.08 inches deep, not enough to take out all the bumps and leave a neat hole with a smooth floor.

—NASA/JPL

---

221 Aquarius 27 **2025 April 02 15:38 UTC**

Ls 065.0° Curiosity Sol 4499
MSD 53768 Perseverance Sol 1464

On this sol in 210 (2004 July 24):

MER-B Sol 178: *Opportunity's* alpha particle X-ray spectrometer examined Diamond Jenness's composition in the abraded area during early morning.

The team decided that the sol would be used to grind into Diamond Jenness again in hopes of deepening the hole. The sequence went extremely well with the rock abrasion tool grinding almost an additional 5 millimeters (about 0.2 inches). The rover then started a Mössbauer spectrometer reading of the deepened hole.

MER-A Sol 199: *Spirit* completed a 6-hour early morning alpha particle X-ray spectrometer reading on Mastodon. After a midday nap to conserve energy, *Spirit* took pictures with the microscopic imager to create a mosaic of the rock abrasion tool hole. *Spirit* then placed the Mössbauer instrument in the hole and began a 20-hour overnight reading.

—NASA/JPL

---

221 Aquarius 28 **2025 April 03 16:18 UTC**

Ls 065.4° Curiosity Sol 4500
MSD 53769 Perseverance Sol 1465

On this sol in 210 (2004 July 25):

MER-B Sol 179: *Opportunity* completed the Mössbauer integration, gathered some remote-sensing data, then positioned the alpha particle X-ray spectrometer on the abraded hole for an early-morning integration at cold temperature on Sol 180. This double integration of the hole (once on Sol 178 at an intermediate depth and then a second one at full depth) will give the science team a unique opportunity to evaluate how the composition changes with depth.

MER-A Sol 200: *Spirit* completed the overnight Mössbauer reading on the rock abrasion tool hole, took a midday nap, stowed the arm, bumped back to take pictures and readings of the hole with the panoramic camera and miniature thermal emission spectrometer, then drove about 52 feet (16 meters). Due to the nature of the terrain, the drive was done in 6-wheel mode to minimize errors (rather than the current standard 5-wheel mode to conserve the aging right front wheel). Engineers carefully targeted *Spirit's* drive to end in a location with favorable tilt to the north to point the solar panels toward the Sun, giving *Spirit* as much power as possible as the Sun hangs low in the sky during Martian southern hemispheric winter.

—NASA/JPL

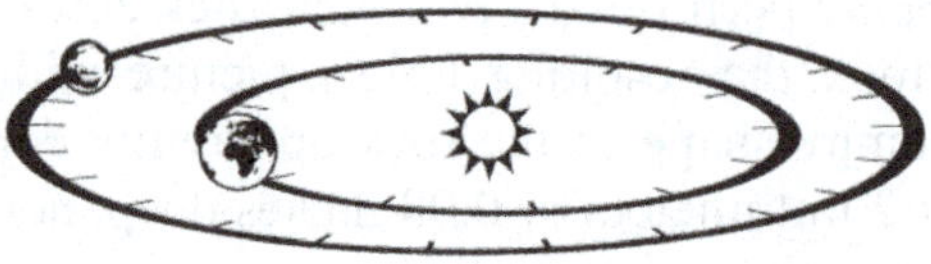

221 KUMBHA

As seen from Mars, the Sun traverses the eastern sector of the constellation Aquarius. In the Greek tradition, the constellation came to be represented as a vase from which a stream poured down to Piscis Austrinus, the Southern Fish. The Sanskrit name of the constellation is Kumbha (कुम्भ), meaning "water-pitcher." Kumbha is a month in the Hindu solar calendar that originally occurred when the Sun passed through the constellation Aquarius as seen from Earth.

The median one-way communications delay during this month is 10 minutes 52 seconds.

← 221 Kumba 01 **2024 October 14 02:06 UTC →**

Ls 065.9° Curiosity Sol 4501
MSD 53770 Perseverance Sol 1466

Approximately the first sol of the unnamed month between November and December on the Heinlein calendar.

On this sol in 194 (1974 June 24):

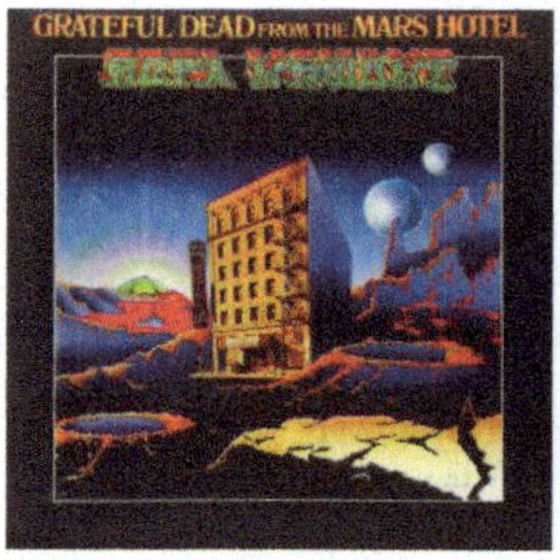

From the Mars Hotel by the American rock band The Grateful Dead was released on vinyl LP. Although no songs on the album were about Mars, its science fantasy cover transported an antique San Francisco edifice to a cratered surface under two moons, with the band members appearing as aliens (or possibly acid trip apparitions) relaxing in front of a television set in one of its rooms.

The Mars Hotel stood at the corner of Fourth and Howard streets in what had become a skid row section of San Francisco, a few blocks from the studio where The Grateful Dead recorded the album. It was demolished in 1980 to make way for construction of the George R. Moscone Convention Center.

On this sol in 198 (1982 January 01):

The song "Melody from Mars" by the British singer-songwriter and rhythm guitarist Ian Gomm was released on his vinyl LP *The Village Voice* (date approximate).

On this sol in 210 (2004 July 26):

MER-B Sol 180: This self-portrait of NASA's Mars Exploration Rover *Opportunity* comes courtesy of the Sun and the rover's front hazard-avoidance camera. The dramatic snapshot of *Opportunity's* shadow was taken as the rover continues to move farther into Endurance Crater.

MER-A Sol 201: *Spirit* was commanded to drive 98 feet (30 meters) across terrain that was pretty steep. *Spirit* accomplished 8.2 feet (2.5 meters) then stopped due to an excessive tilt angle of 25.6 degrees. Engineers had set the maximum tilt angle limit at 25 degrees. *Spirit* did complete pre-drive science observations and post-drive imaging.

—NASA/JPL

221 Kumbha 02 **2025 April 05 17:37 UTC**

Ls 066.3° Curiosity Sol 4502
MSD 53771 Perseverance Sol 1467

On this sol in 210 (2004 July 27):

MER-B Sol 181: A very accurate drive placed the target Mackenzie squarely in *Opportunity's* work volume.

MER-A Sol 202: *Spirit* repeated the drive plan from Sol 201 with the maximum tilt angle set to 32 degrees. This time the rover completed the drive as planned, traveling 83.6 feet (25.5) meters up the hill. *Spirit* then performed post-drive imaging.

—NASA/JPL

221 Kumbha 03 **2025 April 06 18:16 UTC**

Ls 066.8° Curiosity Sol 4503
MSD 53772 Perseverance Sol 1468

On this sol in 185 (1957 July 21):

The CBS Radio Workshop presentation of the radio drama "The Green Hills of Earth" by Robert A. Heinlein was broadcast. A song-writing space engineer is blinded in an accident; later, when nearing the end of his life, he wishes to make a final trip home to see the "green hills of earth." The broadcast included the poet Rhysling's "The Grand Canal" (see 220 Taurus 09).

On this sol in 210 (2004 July 28):

MER-B Sol 182: A two-hour hour rock abrasion tool (RAT) operation at Mackenzie was followed by an observation with the Mössbauer spectrometer. On this sol, *Opportunity* took panoramic camera images during the abrasion tool operation for the first time. The images were normal. Being able to use the panoramic camera and abrasion tool in parallel is one of the items on the "teach your dog new tricks" list, an effort to help the rover multi-task. The rover went into deep sleep this sol.

MER-A Sol 203: A rock outcrop with a view of the surrounding landscape beckons NASA's Mars Exploration Rover *Spirit* on Sol 203 of its journey of exploration on the red planet. This view is a mosaic of images taken by the rover's navigation camera at a position labeled as Site 80, near the top of the West Spur portion of the Columbia Hills. Directly ahead are rock outcrops that scientists will examine for clues that might indicate the presence of water in the past. In the upper right-hand corner is the so-called Sea of Basalt, consisting of lava flows that lapped onto the flanks of the hills. The view is toward the south. The field of view is approximately 170 degrees from right to left and is presented in a cylindrical projection with geometrical seam correction.

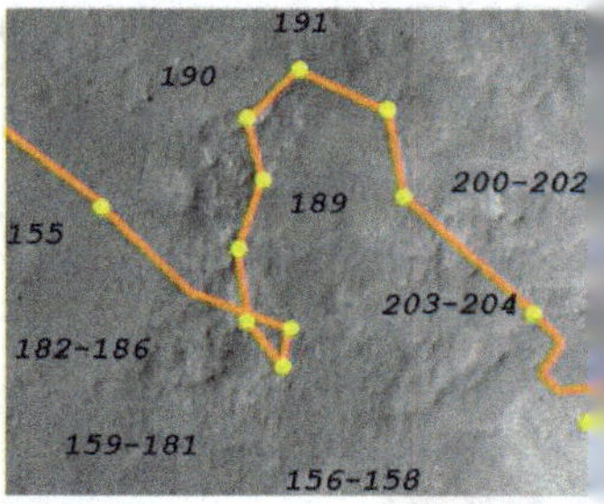

Scientists' hope was to find rock outcropping in this location, but none were found. So the decision was made to continue the drive up the hill to find a better rock outcrop. *Spirit* performed another six-wheel, 62-foot (19-meter) drive. This drive was completed successfully; however, at the end of the drive, *Spirit* drove into a small hollow. As a result, *Spirit* was pitched 15 degrees toward the southwest, and ended up with a southerly tilt.

Mars has seasons like the Earth does, but the seasons are twice as long due to Mars's larger orbit around the Sun. Right now, Mars is approaching northern summer. That also means that it is approaching southern Martian winter at the same time. So, *Spirit* is headed for winter, being 14 degrees south of the equator. Because Martian winter is setting in, solar array energy continues to be a concern for *Spirit.* If *Spirit* parks with a northerly tilt, the rover will see between 350 and 380 watt-hours of energy, but if *Spirit* stops on flat ground or with a southerly tilt, solar energy is as low as 280 watt-hours. So, engineers make a concerted effort to find the north-facing islands along *Spirit's* path.

—NASA/JPL

221 Kumbha 04 **2025 April 07 18:56 UTC**

Ls 067.2° Curiosity Sol 4504

MSD 53773 Perseverance Sol 1469

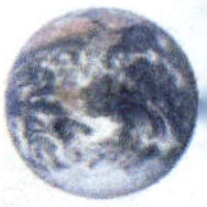

On this sol in 210 (2004 July 29):

MER-B Sol 183: *Opportunity* completed the Mössbauer observation of the RAT hole at Mackenzie, then placed the alpha particle X-ray spectrometer on the hole for an observation to start at 4 a.m. the next morning. (This instrument works best when very cold).

MER-A Sol 204: Planning for this sol was very exciting due to the late downlink of information from Sol 203 (210 Kumbha 03). Very late in the planning cycle, available power on Sol 204 was reduced from 370 watt-hours to 288 watt-hours. Pre-drive observations were cut back to 17 minutes, during which the motors were heated for driving. *Spirit* drove only 0.82 feet (0.25 meters). Because the drive was so short, the power situation is not as bad as it could have been.

Total odometry after Sol 204 is 2.21 miles (3,565.57 meters). Total elevation above the plains of Gusev Crater is estimated to be 30 feet (9 meters).

—NASA/JPL

221 Kumbha 05 **2025 April 08 19:36 UTC**

Ls 067.7° Curiosity Sol 4505
MSD 53774 Perseverance Sol 1470

On this sol in 210 (2004 July 30):

MER-B Sol 184: *Opportunity* took microscopic imager pictures of the Mackenzie RAT hole, stowed the arm, then backed up to observe the hole with the panoramic camera and the miniature thermal emission spectrometer. The rover drove forward roughly 8 meters (26 feet) to Inuvik, using visual odometry to gauge the amount of slip. The drive left *Opportunity* to the right of the intended location because the rover slipped towards the fall line of the crater, causing the vehicle to effectively arc to the right. Deep sleep was invoked.

MER-A Sol 205: *Spirit* attempted to reach Clovis by climbing out of the sandy hollow in which it was sitting. Unfortunately, on a slope of more than 20 degrees, slippage caused *Spirit* to dance around the outcrop. The drive was finally cut off by a time-of-day limit on rover mobility.

—NASA/JPL

---

221 Kumbha 06 **2025 April 09 20:15 UTC**

Ls 068.1° Curiosity Sol 4506
MSD 53775 Perseverance Sol 1471

On this sol in 151 (1893 August 12):

Howard Irving Smith was born; American actor; starred in The Mercury Theatre on the Air's radio play "Invasion from Mars" (1938).

On this sol in 210 (2004 July 31):

MER-B Sol 185: *Opportunity* performed a short maneuver to get Tuktoyuktuk into the arm's work volume. Slippage was greater than expected during the uphill part of that move, so *Opportunity* ended up with only the upper part of the target in the work volume. That turned out to be good enough to perform a full set of arm work, which is planned for Sols 186 through 188 (210 Kumbha 07 through 09). The rover took panoramic camera images of the area between Inuvik and the Arctic Islands for the purposes of evaluating that drive. It turned the inertial measurement unit on again during the afternoon communications relay. This is another item on the "new tricks" list that, if successful, will allow rover planners to turn the vehicle during communication passes to optimize the data return. The rover again used deep sleep.

Opportunity is completing an intensive survey of the Karatepe region that began 50 sols ago when the rover first ventured into Endurance Crater. The rover currently sits about 20 meters (about 66 feet) inside the stadium-sized crater. The investigation at an area dubbed Inuvik at a target called Tuktoyuktuk (named for a small village in the Canadian arctic) will likely be the rover's last in this region. The rover planning team is contemplating the next traverse which will move *Opportunity* around the interior of the crater, first to some outcroppings dubbed the Arctic Islands, then possibly to Burns Cliff, roughly 80 meters (about 262 feet) from the rover's current position. *Opportunity* continues to perform very well, a testament to all those who worked so hard to get it to Mars and to those who operate it daily.

Some concerns that are being addressed are slippage, an error message from the microscopic imager and pointing errors with the front hazard-avoidance cameras.

The drive on Sol 185 included a short backup, during which the rover experienced a 40 percent slip. Typical slips when driving uphill have been in the 15- to 20-percent range. More evaluation of what happened on this and other drives will be needed before any general conclusions can be made about traversability in this region. The overall slope in this area is 15 degrees, which is 10 degrees below the general threshold of concern for rocky terrain.

There have been four instances of a warning message in the last ten sols that indicate a problem getting data from the microscopic imager. The messages indicate that the data was corrupted, and that a retry was necessary to receive the data without error. In all cases, the retry succeeded in transferring the data. This problem has not been seen before on either vehicle.

The new front hazard-avoidance camera models may need some more tweaking. Pointing errors were greater than expected on two recent placements of the instrument deployment device (robotic arm). The error is such that rover planners can still confidently place the instruments, provided that a 2-centimeter (0.8-inch) offset can be safely tolerated. If more precision is needed, planners must first use the microscopic imager to survey the target, then wait one sol before placing any instruments.

MER-A Sol 206: The plan for this sol was designed to accommodate up to a 50 percent slip and still reach the outcrop target. However, due to challenging terrain near Clovis, *Spirit* again did not end up exactly where scientists and engineers wanted it to go. For part of its traverse, *Spirit* slipped about 125 percent, actually losing ground in its attempt to move uphill.

Late in the sol, internal software experienced a timing problem in which two instrument-related commands were given at nearly the same time, temporarily precluding further operation of the miniature thermal emission spectrometer and camera mast on *Spirit.*

—NASA/JPL

221 Kumbha 07 **2025 April 10 20:55 UTC**

Ls 068.6° Curiosity Sol 4507

MSD 53776 Perseverance Sol 1472

Earth departure opportunity to Mars via an EVM trajectory.

Mars arrival opportunity from Earth via a Type I trajectory.

On this sol in 152 (1895 July 01):

Part 3 of *Mars* by Percival Lowell appeared in *The Atlantic* (date approximate).

On this sol in 184 (1955 September 05):

The *X Minus One* presentation of the radio drama "The Green Hills of Earth" by Robert A. Heinlein was broadcast. A song-writing space engineer is blinded in an accident; later, when nearing the end of his life, he wishes to make a final trip home to see the "green hills of earth." The broadcast included the poet Rhysling's "The Grand Canal" (see 220 Taurus 09).

On this sol in 210 (2004 August 01):

MER-B Sol 186: After a night of deep sleep, *Opportunity* started the sol with imaging of the sky in search of clouds and using its miniature thermal emission spectrometer for observations of the sky and ground. In the afternoon, the rover took microscopic images of a target called Tuktoyuktuk, then used its rock abrasion tool to gnaw a hole 7.7 millimeters (0.3 inch) deep into that target. The robotic arm moved the alpha particle X-ray spectrometer into position for reading of the composition of the freshly exposed interior of the rock.

Opportunity has driven about 20 meters (66 feet) into Endurance Crater, examining increasingly older layers of bedrock as it advances. If assessments of traversability continue giving positive indications, the rover team plans next to send *Opportunity* counterclockwise across the inner slope of the crater to study possible targets of dune tendrils, boulders and the base of a cliff.

Four times in the past two weeks, *Opportunity* has sent error messages while successfully taking pictures with its microscopic imager. One theory for the cause is degradation of flexible cabling that runs down the rover's robotic arm to the instrument. As a precaution while undertaking further analysis, the rover team is treating use of the arm as a consumable resource, with cable wear each time the arm is moved decreasing the possible number of future microscopic images.

"We are being very conservative about this because we certainly don't want to do anything to jeopardize the instruments," said Dr. Ken Herkenhoff of the U.S. Geological Survey Astrogeology Team, Flagstaff, Ariz., lead scientist for both rovers' microscopic imagers. "We are running more diagnostics that we hope will identify the problem. There are potential explanations that would mean we do not have to treat arm use as a consumable."

Erickson said, "We will no doubt have more issues with them in the future. We'll do everything we can to milk the most value out of them while they are usable, but they won't last forever."

MER-A Sol 207: While *Spirit* was executing commands, a semiconductor component failed to power on as intended. The component, a programmable gate array, directly affects usability of the rover's three spectrometer instruments. Subsequent commands for using the miniature thermal emission spectrometer in that day's sequence resulted in repeated error messages.

Engineers on the Mars Exploration Rover team at NASA's Jet Propulsion Laboratory, Pasadena, California, have determined the most likely cause is a timing issue of one instruction reaching the gate array microseconds before another that was intended to precede it. If that diagnosis is confirmed, a repeat could be avoided by inserting a delay between commands that might reproduce the problem, engineers expect. Until then, the rover science team's daily choices for how to use *Spirit* do not include using the miniature thermal emission spectrometer, the Mössbauer spectrometer or the alpha particle X-ray spectrometer.

"While we're being very cautious in how we operate today and tomorrow, we expect to verify the problem and resolve this issue with a relatively easy workaround," said JPL's Jim Erickson, project manager for the twin rovers.

Spirit has driven to a bedrock exposure near the top of a spur of the Columbia Hills. The location sits about nine meters (30 feet) above a plain that the rover crossed for months to get from its landing site to the hills. Planners intend for *Spirit* to spend more than a week at this site, inspecting the rock exposure, dubbed Clovis, and recording the panoramic scene from this viewpoint.

—NASA/JPL

221 Kumbha 08 **2025 April 11 21:34 UTC**

Ls 069.0° Curiosity Sol 4508

MSD 53777 Perseverance Sol 1473

On this sol in 210 (2004 August 02):

MER-B Sol 187: *Opportunity* woke for an early morning *Mars Odyssey* communications relay session. After that, the rover started the alpha particle X-ray spectrometer reading, which lasted until 9:00 a.m. local solar time. *Opportunity* then took a long nap as the uplink command window was delayed until 1:00 p.m. local solar time due to launch of NASAs *Messenger* mission to Mercury. In the afternoon, *Opportunity* rotated its tool turret from the alpha particle X-ray spectrometer to the Mössbauer spectrometer and acquired a 7-hour Mössbauer reading before beginning deep sleep.

The rover team is addressing some concerns about rover slippage and about error messages from the microscopic imager. An uphill driving test is planned for Sol 188 to gain better understanding of a 40-percent slip observed in a repositioning maneuver on Sol 185 (210 Kumbha 06). This will aid planning for a potential drive clockwise across the inner slope of the crater toward rocks called the Arctic Islands and the base of Burns Cliff. Diagnostic work is also planned for Sol 188 (210 Kumbha 09) about the error messages generated during use of the microscopic imager.

MER-A Sol 208: While *Spirit's* timing issue was being analyzed, engineers decided not to use the mast, panoramic cameras, navigation cameras, Mössbauer spectrometer, alpha particle X-ray spectrometer, or the miniature thermal emission spectrometer. On the bright side, since the problem did not affect communications, a communications experiment with the European Space Agency's *Mars Express* orbiter was successfully conducted in the early morning hours of Sol 208.

By this sol, *Spirit's* mast had been declared usable. Operators commanded *Spirit* to drive 7.5 meters (25.6 feet) to reach a rock outcrop called Clovis, using a route avoiding the steepest terrain that had created problems for the rover in earlier sols.

Over the last few sols, *Spirit* struggled mightily to Clovis, overcoming the challenge of rough, steep terrain and subsequent backsliding. The site is near the crest of the West Spur of Columbia Hills.

Spirit is examining Clovis. This outcrop will likely be the subject of *Spirit's* most intensive investigation to date.

—NASA/JPL

221 Kumbha 09 **2025 April 12 22:14 UTC**

Ls 069.4° Curiosity Sol 4509
MSD 53778 Perseverance Sol 1474

On this sol in 173 (1935 January 01):

Part 3 of *Swords of Mars* by Edgar Rice Burroughs appeared in *The Blue Book Magazine* (date approximate).

Interior art by Robert Fink.

On this sol in 210 (2004 August 03):

MER-B Sol 188: The sol was devoted to finishing an examination of a target patch called Tuktoyuktuk where the rover's rock abrasion tool had ground the surface coating off of a rock called Inuvik. *Opportunity* then drove partly up-slope and partly cross-slope as both a mobility test and the start of a traverse to the next target. It slipped down-slope about as much as expected, a good result. The rover's current terrain consists of rocky plates lightly covered with sand and soil, plus some deeper sandy patches between the plates. The sandy patches result in more slip and sometimes cause the vehicle to yaw a little (as more slip on one side of the vehicle than the other causes it to turn). The overall tilt of the rover is about 18 degrees.

MER-A Sol 209: *Spirit* experienced an unexpected reboot of the flight software. This incident was not a threat to the spacecraft. It is a known bug in the system that the rover team is working around.

—NASA/JPL

221 Kumbha 10 **2025 April 13 22:53 UTC**

Ls 069.9° Curiosity Sol 4510
MSD 53779 Perseverance Sol 1475

On this sol in 210 (2004 August 04):

MER-B Sol 189: *Opportunity* drove about 4 meters (13 feet) eastward across the inner slope of the crater. The drive went well despite substantial down-slope slip. Slippage averaged about 33 percent, with a peak of about 56 percent on one half-meter (1.6-foot) drive segment, but the rover team expected that and compensated in advance for it. The team then asked the rover to conduct a series of turns in place during the communications relay pass with the *Mars Odyssey* orbiter to optimize the communications link. The idea was to keep *Mars Odyssey* in the sweet spot of the rover's ultra-high-frequency antenna pattern as the orbiter swept across the sky. The total data return was about 135 megabits. The best possible return predicted by models if the rover had just sat in one orientation was about 115 megabits.

The next target the scientists would like the rover to approach, Axel Heiberg, is a rocky outcrop about 18 meters (59 feet) away to the east and a bit deeper in the crater.

MER-A Sol 210: *Spirit* drove up steep terrain to reach the exact spot on Clovis for work with the science instruments at the end of the robotic arm.

—NASA/JPL

---

221 Kumbha 11 | **2025 April 14 23:33 UTC**

Ls 070.3° | Curiosity Sol 4511

MSD 53780 | Perseverance Sol 1476

On this sol in 210 (2004 August 05):

MER-B Sol 190: *Opportunity* continues its voyage farther into Endurance Crater with a near-term drive goal of a rock outcrop dubbed Axel Heiberg, and a possible later destination at the foot of Burns Cliff on the south side of the crater.

Opportunity completed a 3.4-meter (about 11-foot) drive towards Axel Heiberg. The slope was steady at about 17 degrees and slippage during the drive was about 16 percent, as predicted. The rover took images for use in planning future drives and made observations with its miniature thermal emission spectrometer. Controllers employed the microscopic imager to help with diagnosing the cause of error messages from that instrument received last week.

As NASA's Mars Exploration Rover *Opportunity* creeps farther into Endurance Crater, the dune field on the crater floor appears even more dramatic. This approximate true-color panoramic camera image highlights the reddish-colored dust present throughout the scene.

Sinuous tendrils of sand less than 1 meter (3.3 feet) high extend from the main dune field toward the rover. Scientists hope to send the rover down to one of these tendrils in an effort to learn more about the characteristics of the dunes. Dunes are a common feature across the surface of Mars, and knowledge gleaned from investigating the Endurance dunes close-up may apply to similar dunes elsewhere.

Before the rover heads down to the dunes, rover drivers must first establish whether the slippery slope that leads to them is firm enough to ensure a successful drive back out of the crater. Otherwise, such hazards might make the dune field a true sand trap.

—NASA/JPL

221 Kumbha 12 **2025 April 16 00:13 UTC**

Ls 070.8° Curiosity Sol 4512
MSD 53781 Perseverance Sol 1477

Aphelion occurs at 21:36 AMT (August 06 22:27 UTC).

On this sol in 210 (2004 August 06):

MER-B Sol 191: *Opportunity* successfully drove another 5 meters (about 16.4 feet) closer to Axel Heiberg, leaving about 5 meters (about 16.4 feet) to go. The drive included a short backup at the end to check for uphill-drive slippage, which was within acceptable limits. Deep sleep was used overnight.

The team continues to acquire microscopic imager diagnostic images at different times of day to see if temperature might be a contributing factor to the errors seen from that instrument last week. So far, no more errors have occurred.

—NASA/JPL

221 Kumbha 13 **2025 April 17 00:52 UTC**

Ls 071.2° Curiosity Sol 4513
MSD 53782 Perseverance Sol 1478

On this sol in 210 (2004 August 07):

MER-B Sol 192: *Opportunity* drove down slope in Endurance Crater to reach a rock dubbed Axel Heiberg. The rover arrived in position to approach a particular point for work in this area with its instrument deployment device (the rover arm). Favorable geometry for an overnight communications pass with *Mars Odyssey* motivated the team to keep the rover out of deep sleep mode and take advantage of the pass. About 115 megabits of data were returned in this overnight pass.

—NASA/JPL

221 Kumbha 14 **2025 April 18 01:32 UTC**

Ls 071.7° Curiosity Sol 4514
MSD 52783 Perseverance Sol 1479

On this sol in 210 (2004 August 08):

MER-B Sol 193: *Opportunity* started the sol with some cloud imaging at about 8:45 a.m., local solar time. This required some heating of the camera mast motors and bearings. The observations were acquired, though one of the heaters apparently did not heat as planned. Engineers believe a thermostat controlling that heater had already opened. This rendered the heating circuit inoperable so that even though the heater switch was commanded on and off correctly, the heater itself never got powered. This probably resulted in use of the mast actuator at lower-than-intended temperatures. Rover team members are investigating this, and in the meantime they will not command the rover to perform mast activities at that time of morning.

After the early morning activities, *Opportunity* was commanded to approach a target on Axel Heiberg for grinding with the rock abrasion tool. The drive was designed and executed to compensate for slip, and the result was very precise. The rover also made additional remote-sensing observations, then it went into a deep sleep for the night to save energy.

—NASA/JPL

221 Kumbha 15 **2025 April 19 02:11 UTC**

Ls 072.1° Curiosity Sol 4515

MSD 52784 Perseverance Sol 1480

On this sol in 210 (2004 August 09):

MER-B Sol 194: Opportunity took microscopic imager pictures of a spot on Axel Heiberg, and then performed a grind with the rock abrasion tool to get access to subsurface chemistry. The grind went well, but the targeting was a little off (the hole was about 6 centimeters - about 2.4 inches - to the left of the intended target.) After some investigation it was determined that there is an error in the way one of the ground tools represented the commanded position. This error has existed previously, but the team has never detected it to be this large. It is now being fixed. The exact positioning of the rover and the arm, and the nature of the activity all combined to make the error particularly large in this instance. After the grind, the rover placed the alpha particle X-ray spectrometer on the hole to measure the rock's elemental composition early the next morning.

—NASA/JPL

221 Kumbha 16 **2025 April 20 02:51 UTC**

Ls 072.6° Curiosity Sol 4516

MSD 52785 Perseverance Sol 1481

On this sol in 210 (2004 August 10):

MER-B Sol 195: *Opportunity* acquired post-grind microscopic images and placed the Mössbauer instrument on the hole to take a reading all afternoon. The rover also made remote-sensing observations, including images to help assess where it might drive next.

On this sol in 210 (2004 August 11):

MER-A Sol 216: Between Sols 211 and 216 (210 Kumbha 11 and 16), *Spirit* completed an alpha particle X-ray spectrometer reading of a spot on Clovis called Plano, which had been brushed off using the rock abrasion tool. *Spirit* then placed the rock abrasion tool on Plano again and drilled for 2.5 hours, creating a hole 8.9 millimeters (0.4 inch) deep, which is a new record. *Spirit* also continued a campaign to capture a color 360-degree panoramic camera image from this location.

—NASA/JPL

221 Kumbha 17 **2025 April 21 03:31 UTC**

Ls 073.0° Curiosity Sol 4517

MSD 52786 Perseverance Sol 1482

On this sol in 148 (1888 January 01):

Chesley Knight Bonestell, Jr. was born; American painter, designer and illustrator. His paintings inspired the American space program, and they have been (and remain) influential in science fiction art and illustration. A pioneering creator of astronomical art, Bonestell has been dubbed the "Father of Modern Space Art."

—Wikipedia

His works included numerous paintings of the Martian globe and the planet's landscapes.

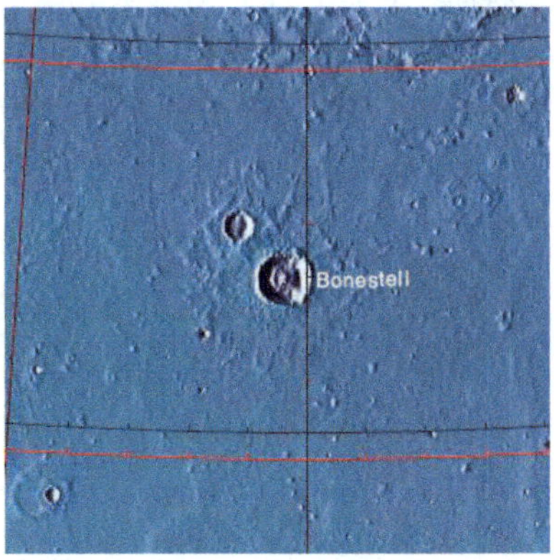

42.4-kilometer diameter crater located at 42.3°N, 329.5°E

On this sol in 176 (1940 September 01):

"The Tyrant of Mars" by Frederic Arnold Kummer Jr. appeared in *Thrilling Wonder Stories* (date approximate). A scientist's strategy sunders the shackles of a crimson world in bondage!

Interior art by Leo Morey.

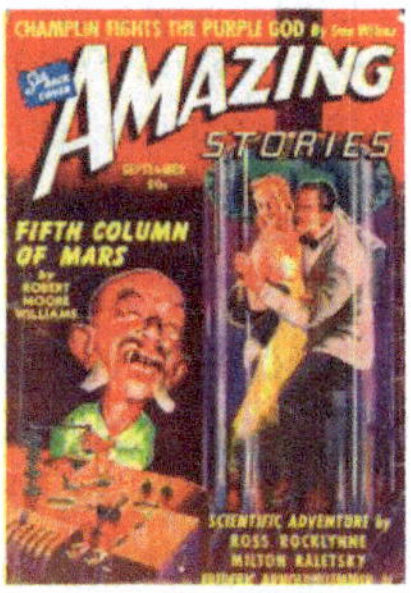

"Fifth Column of Mars" by Robert Moore Williams appeared in *Amazing Stories* (date approximate). How could a U.S. secret service man suspect the fifth column he trailed would lead him to Mars?

Cover art by Robert Fuqua.

On this sol in 210 (2004 August 11):

MER-B Sol 196: *Opportunity* completed an overnight reading with the Mössbauer spectrometer on a hole into Axel Heiberg where the rock abrasion tool gnawed off the rock's outer surface on Sol 193 (210 Kumbha 14). Then the rover bumped back about half a meter (about 1.6 feet) to position itself for reaching an interesting vein feature. After the bump, *Opportunity* made observations of the abraded hole with its miniature thermal emission spectrometer and panoramic camera data to complete the remote sensing of that target.

Opportunity imaged Martian clouds.

On this sol in 210 (2004 August 12):

MER-A Sol 217: *Spirit* took microscopic images of the rock abrasion tool hole Plano on Clovis, and then placed the alpha particle X-ray spectrometer in the hole for an early morning observation.

—NASA/JPL

221 Kumbha 18 **2025 April 22 04:10 UTC**

Ls 073.5° Curiosity Sol 4518

MSD 52787 Perseverance Sol 1483

On this sol in 210 (2004 August 12):

MER-B Sol 197: *Opportunity* completed a microscopic imager mosaic of the vein feature called Sermilik. The rover also acquired panoramic camera and miniature thermal emission spectrometer data of sand spots to identify future targets for the rover arm. Overnight deep sleep was used to conserve power.

On this sol in 210 (2004 August 13):

MER-A Sol 218: *Spirit* successfully performed a couple of communications tests with the European Space Agency's *Mars Express* orbiter last week. The tests demonstrated the two spacecraft's ability to work together to transmit data collected by the rovers to Earth via the *Mars Express* communications relay. NASA's *Mars Odyssey* and *Mars Global Surveyor* orbiters also have this capability. More than 85 percent of the data from the rovers has been transmitted to Earth via the *Mars Odyssey* orbiter.

Spirit continued work over the past nine sols at a rock called Clovis. The rover used its rock abrasion tool, microscopic imager, alpha particle X-ray spectrometer, and Mössbauer spectrometer to probe deeper into the history of this rock. Clovis is the most altered rock encountered by *Spirit* to date. It is part of a rock outcrop located on the West Spur of the Columbia Hills, roughly 55 meters (180 feet) higher than *Spirit's* landing site about 3 kilometers (2 miles) away.

Spirit placed the Mössbauer spectrometer in the rock abrasion tool hole and started a 48-hour observation. This is a longer than normal integration time, with a goal of resolving in more detail the makeup of this highly altered rock.

—NASA/JPL

---

221 Kumbha 19 **2025 April 23 04:50 UTC**
Ls 073.9° Curiosity Sol 4519
MSD 52788 Perseverance Sol 1484

On this sol in 210 (2004 August 13):

MER-B Sol 198: *Opportunity* awoke from deep sleep and used heaters to warm the panoramic camera mast assembly in preparation for morning cloud, sky and ground imaging and miniature thermal emission spectrometer observations. It made a daytime Mössbauer inspection of a 3-centimeter (1.2-inch) chuck of vein material that was apparently broken off from the vein when the rover backed up. This was followed by a tool change to the alpha particle X-ray spectrometer before sleep. *Opportunity* awoke for an early morning *Mars Odyssey* communications pass and turned on the alpha particle X-ray spectrometer for a nearly 6-hour integration.

—NASA/JPL

---

221 Kumbha 20 **2025 April 24 05:29 UTC**
Ls 074.4° Curiosity Sol 4520
MSD 52789 Perseverance Sol 1485

On this sol in 184 (1955 September 20):

The *X Minus One* presentation of the radio drama "The Green Hills of Earth" by Robert A. Heinlein was broadcast. A song-writing space engineer is blinded in an accident; later, when nearing the end of his life, he wishes to make a final trip home to see the "green hills of earth." The broadcast included the poet Rhysling's "The Grand Canal" (see 220 Taurus 09).

On this sol in 210 (2004 August 14):

MER-B Sol 199: *Opportunity's* planned abrading of a target called Jiffypop failed. The preparatory seek-scan process successfully found the rock surface, but a motor stall prevented any further activity by the rock abrasion tool. Planned microscopic imager pictures of the target area and remote sensing were acquired successfully.

—NASA/JPL

---

221 Kumbha 21 **2025 April 25 06:09 UTC**
Ls 074.8° Curiosity Sol 4521
MSD 52790 Perseverance Sol 1486

On this sol in 153 (1897 June 01):

Part 9 of *The Martian* by George du Maurier appeared in *Harper's New Monthly Magazine* (date approximate).

On this sol in 174 (1936 December 01):

The Master of Mars by James Edison Northfield appeared in the only issue of *Flash Gordon Strange Adventure Magazine* (date approximate).

Cover and interior art by Fred Meagher.

"The Brain Stealers of Mars" by John W. Campbell, Jr. appeared in *Thrilling Wonder Stories* (date approximate). Penton and Blake find Mars a pleasant place to be—until they run afoul of the thushol, which can imitate anything!

Interior art by Alex Schomburg.

"Time Measures on Mars" by Robert G. Aitken, Director Emeritus of Lick Observatory, was published as "Astronomical Society of the Pacific Leaflet 95" (date approximate). The paper described in detail the first civil calendar for Mars, lacking only an epoch by which to count the marsrevs. Aitken divided the marsrev into four equal "seasons," each of which had three "quarters" of 42 sols and one of 41 sols. A mid-marsrev sol was to be inserted in marsrevs divisible by 10. On Aitken's calendar the date was Spring Quarter 4, Sol 35.

"The Space Marines and the Slavers" by Bob Olsen appeared in *Amazing Stories* (date approximate). Sequel to "Captain Brink of the Space Marines." The Martian space pirate Zurek has just raided the settlement on Ganymede, killing the old and very young, and carrying away the others as slaves. Among those kidnapped is Ingeborg Andersen, the sweetheart of Dan, the cadet member of Captain Brink's team on the *Hyperion.*

—Bleiler and Bleiler, 1998

Cover and interior art by Leo Morey.

On this sol in 210 (2004 August 15):

MER-B Sol 200: *Opportunity* was commanded to perform some remote sensing and some rock abrasion tool diagnostics in response to an activity that faulted out on Sol 199. During these diagnostics, the tool failed to respond as desired to a command to calibrate the grind motor. Analysis of this event suggests that there is a piece of debris (probably a rocky chunk of Mars) trapped between the grind bit and the brush bit. The rover team believes that it can be freed by turning the bits in reverse, but they are still evaluating the best approach to remedy the situation. There are several options available. The team decided to continue the investigation of this anomaly while pressing on with other objectives.

—NASA/JPL

221 Kumbha 22 **2025 April 26 06:48 UTC**

Ls 075.3° Curiosity Sol 4522
MSD 52791 Perseverance Sol 1487

Mars arrival opportunities from Earth via a Type I trajectory and an EEM trajectory.

Earth arrival opportunity from Mars via a Type II trajectory.

On this sol in 181 (1950 February 01):

“The Long Weekend” by Ray Bradbury appeared in *Argosy.* The story was originally published as “The Million Year Picnic,” and appeared as the final vignette in *The Martian Chronicles.*

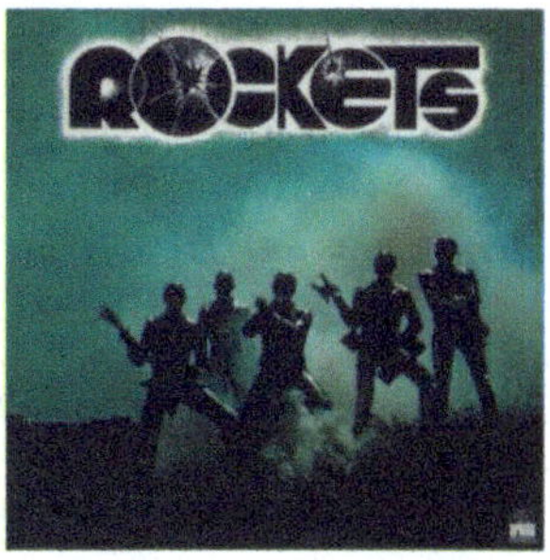

On this sol in 195 (1976 June 01):

The song “Balade sur Mars” by the Italian rock band The Rockets was released on their self-titled debut studio vinyl LP (date approximate).

On this sol in 210 (2004 August 16):

MER-B Sol 201: *Opportunity* was commanded to stow its arm and drive to a position about 12 meters (about 39 feet) clockwise around the crater. The intent is to head towards a dune tendril that reaches out of the bottom of the crater and may be accessible without having to drive into terrain that is too sandy for the rover to safely traverse. The drive went very well, and the rover ended up in the expected place.

—NASA/JPL

221 Kumbha 23 **2025 April 27 07:28 UTC**

Ls 075.7° Curiosity Sol 4523
MSD 52792 Perseverance Sol 1488

On this sol in 154 (1899 April 22):

Byron Haskin was born; American film director; directed the motion pictures *The War of the Worlds* (1953), *Conquest of Space* (1955) , and *Robinson Crusoe on Mars* (1964), as well as “The Invisible Enemy” episode of *The Outer Limits* (1964).

On this sol in 172 (1933 March 01):

“Salvage in Space” by Jack Williamson appeared in *Astounding Stories of Super-Science* (date approximate). To Thad Allen, meteor miner, comes the dangerous bonanza of a derelict rocket-flier manned by death invisible.

—Goodreads.com

Cover art by H.W. Wesso. Interior art by Marco

Marchioni.

"Dweller in Martian Depths" (later reprinted as "The Dweller in the Gulf") by Clark Ashton Smith appeared in *Wonder Stories* (date approximate). Three Earthmen, hard-bitten prospectors, are exploring the Chaur desert region for gold, which is said to be plentiful. Sighting an approaching sandstorm, they take refuge in a nearby cavern mouth, which as they soon see, is really the entrance to an enormous pit, around the side of which winds a pathway.

—Bleiler and Bleiler, 1998

Interior art by Frank R. Paul.

"Invasion" by Murray Leinster appeared in *Astounding Stories of Super-Science* (date approximate). The whole fighting fleet of the United Nations is caught in Kreynborg's marvelous, unique trap.

Interior art by H.W. Wesso.

On this sol in 183 (1953 November 07):

Rod Brown of the Rocket Rangers Season 1 Episode 30 "The Robot Robber of Deimos" airdate. The Rocket Rangers are called into action after a humanoid robot is used to rob a Martian bank.

—International Movie Database

On this sol in 210 (2004 August 17):

MER-B Sol 202: *Opportunity* was commanded to proceed a little ways downslope. Team members were not able to command the drive the rover as far as they might have liked because they did not get all the data they hoped to get in the afternoon downlink pass on Sol 201. The terrain around the rover is heavily coated with sand and dust, so each traverse requires careful evaluation to make sure there is enough rock material to drive on with confidence. From the images available, the team determined it could safely command only about a 1-meter (3.3-foot) drive. This drive proceeded as expected. At the end of the drive, panoramic camera images were acquired directly in front of the rover and out to the dune tendril. These images will be used to assess traversability to this sandy feature.

On this sol in 210 (2004 August 18):

MER-A Sol 223: During sols 219 through 223 (210 Kumbha 19 through 23), *Spirit* completed science observations at the Clovis rock outcrop. So, *Spirit* packed up and slowly moved on. Winter is approaching and temperatures continue to drop. Power is always a major concern as available energy fluctuates between 300 and 400 watt-hours per sol, but *Spirit* continues the quest, climbing ever higher into the Columbia Hills. *Spirit* has climbed more than 13 meters (43 feet) in elevation from Hank's Hollow, at the base of Columbia Hills, but currently is at an elevation of 37 meters (121 feet) above its landing site on the plains of Gusev Crater.

This stunning image mosaic of the Columbia Hills is the first 360-degree panorama taken since the Mars Exploration Rover *Spirit* arrived at the hills over a month ago. The rover has been busy studying the rocks here, which show evidence of past alteration by water. The dark patch of soil to the right is the spot where *Spirit* stopped for engineering work on its right front wheel. *Spirit's* tracks can be followed from there all the way back to Bonneville Crater and the original landing site, more than 3 kilometers (1.86 miles) away.

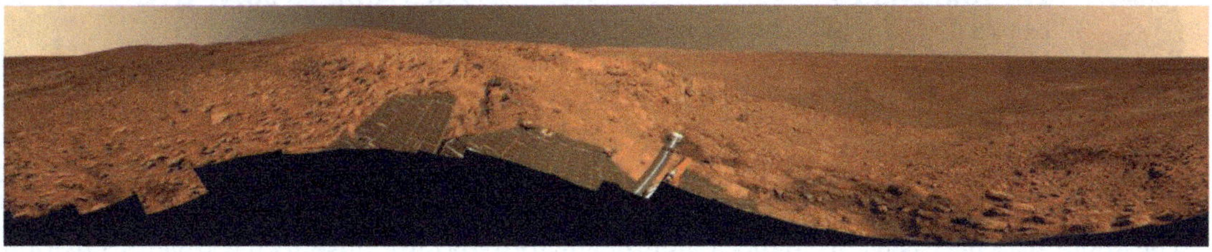

This approximate true-color image, nicknamed the Cahokia panorama after the Native American archaeological site near St. Louis, was acquired between Sols 213 to 223 (210 Kumbha 13 to 23). The panorama consists of 470 images acquired through six panoramic camera filters (750 to 480 nanometers). It took until the week of Sol 237 (210 Pisces 10) to downlink all the data back to Earth. Several more weeks of image processing and geometric mapping by team members at Jet Propulsion Laboratory and Cornell University to stitch all the images together into this spectacular mosaic.

—NASA/JPL

221 Kumbha 24 **2025 April 28 08:08 UTC**

Ls 076.2° Curiosity Sol 4524

MSD 52793 Perseverance Sol 1489

On this sol in 184 (1955 September 26):

Journey into Space: The World in Peril, Episode 1 airdate. 15 April 1972. *Discovery* and two freighters return to the Moon from the disastrous Mars expedition. The crew are subjected to intense questioning about their experiences on Mars, and Lemmy discovers that *Discovery* is being readied for take off. Back on Earth, they visit an astronomical observer who has spotted suspicious objects in the heavens.

—Wikipedia

On this sol in 210 (2004 August 18):

MER-B Sol 203: The team decided to scratch the approach to the dune tendril and, instead, headed *Opportunity* back towards Axel Heiberg and another target named Ellesmere for some soil observations. The terrain between the rover and the dune tendril did not present clear evidence of rocky plates to give the rover sufficient traction. Rather than spend more time in an attempt to scout further for an approach path, the decision was made to abandon the quest for the dune tendril. A drive of approximately 14 meters (46 feet) positioned the rover where it will be able to zero in on Ellesmere next. There was an apparent combination of slip or induced heading change, or both, due to the sandy terrain, which resulted in the rover ending up about 3 meters (about 10 feet) farther left than expected. This also caused *Opportunity* to unintentionally run over a patch of fine soil with some small dune-like ripples in it. The team will be assessing this traverse error, but it is par for the course when driving this far on such sandy, sloped terrain.

On this sol in 210 (2004 August 19):

MER-A Sol 224: Operators became concerned that *Spirit's* batteries might be entering a very low state of charge, so *Spirit* shut down to charge the batteries.

—NASA/JPL

221 Kumbha 25 **2025 April 29 08:47 UTC**
Ls 076.6° Curiosity Sol 4525
MSD 52794 Perseverance Sol 1490

On this sol in 210 (2004 August 19):

MER-B Sol 204: The sol was planned as a rather circuitous 6-meter (about 20-foot) traverse for *Opportunity* to the vicinity of a target called Shag on one side of a rock called Ellesmere. The route was necessary to avoid a significant rock hazard close to the rover's position. Unfortunately, due to the steep slopes and lack of traction when driving in this terrain, the rover experienced up to 50 percent slip during parts of its traverse. It ended up more than 50 centimeters (about 20 inches) downslope from the planned final position. This left the rover close to the edge of its safe terrain zone.

On this sol in 210 (2004 August 20):

MER-A Sol 225: *Spirit* awoke at 11:20 a.m., Gusev local solar time and stayed awake well past the afternoon communication session with NASA's *Mars Odyssey* orbiter without taking a nap. *Spirit* hasn't done that in a long time. The rover extended its arm and used its rock abrasion tool to brush seven circular patches on Clovis. After the first five brushings, the arm was moved out of the way and an image was taken of the circles. These five brushing circles resemble the Olympic rings.

After completing the brushing, *Spirit* performed a 30-minute reading with its alpha particle X-ray spectrometer, imaged the seven rings using its microscopic imager, stowed its arm and drove one meter (three feet) backward, farther off of the Clovis outcrop. *Spirit's* sol was far from over. The rover performed post-drive observations with its navigation camera, observations with its miniature thermal emission spectrometer concurrent with the afternoon *Mars Odyssey* pass, and panoramic camera observations before shutting down.

Total odometry after Sol 225 was 3,605 meters (2.24 miles), a Mars record.

—NASA/JPL

221 Kumbha 26 **2025 April 30 09:27 UTC**
Ls 077.1° Curiosity Sol 4526
MSD 52795 Perseverance Sol 1491

On this sol in 210 (2004 August 20):

MER-B Sol 205: The team shifted *Opportunity's* objective from Shag to another target, Auk, on the other side of Ellesmere. Auk, though farther from the rover's current position, was of higher scientific priority, in safer terrain, and more accessible to the rover arm. To avoid the significant slip observed during turns in place, the traverse was planned as a tight-radius turn covering 1.6 meters (5.2 feet). Mindful of the uncertainties inherent in navigating in this terrain, planners designed the traverse to cover only a portion of the total distance to Auk. This proved to be prudent, since the rover again ended up slipping more than 50 percent during most of its drive, with little progress away from dangerous terrain. On the bright side, analysis of the drive indicated that the rover was getting better traction during its last moves.

The route charted to enter the crater was a bit circuitous, but well worth the extra care engineers took to ensure the rover's safety. On Sol 94, *Opportunity* sat on the edge of this impressive, football field-sized crater while rover team members assessed the scene. After traversing around the Karatepe region and past Burns Cliff, the rover engineering team assessed the possibility of entering the crater.

Careful analysis of the angles *Opportunity* would face, including testing an Earth-bound model on simulated Martian terrain, led the team to decide against entering the crater at that particular place. *Opportunity* then backed up before finally dipping into the crater on its 130th sol (210 Makara 07).

The rover has since made its way down the crater's inner slope, grinding, trenching and examining fascinating rocks and soil targets along the way. The rover nearly made it to the intriguing dunes at the bottom of the crater, but when it got close, the terrain did not look safe enough to cross.

On this sol in 210 (2004 August 21):

MER-A Sol 226: *Spirit* stopped at a rock dubbed Ebenezer for several sols of intense science. Ebenezer is roughly 8 meters (26 feet) from Clovis.

While at Ebenezer, *Spirit* was facing south of east, with its nose pitched up 21 degrees. This orientation was very favorable from a power perspective since the sun tracks to the north. *Spirit's* daily solar energy input increased about 10 percent as a result. Spirit also had a great view of the Gusev plain from this location.

—NASA/JPL

221 Kumbha 27 **2025 May 01 10:06 UTC**

Ls 077.5° Curiosity Sol 4527

MSD 52796 Perseverance Sol 1492

As seen from Mars, the Sun enters the constellation Pisces.

On this sol in 210 (2004 August 21):

MER-B Sol 206: *Opportunity* was commanded to perform a drive to turn away from its cross-slope orientation and move upslope toward Auk. The drive succeeded. After the slips of the Sol 205 traverse, this traverse managed nearly all of the desired yaw response to get the rover pointed uphill and then found good traction to deliver the rover more than a meter (3.3 feet) farther upslope. Serendipitously, the rock directly in front of the rover at the end of the drive proved to be so interesting to the science team that efforts were redirected to study it. The rock was dubbed Escher.

On this sol in 210 (2004 August 22):

Mars Express imaged Phobos.

MER-A Sol 227: *Spirit* continued its examination of Ebenezer.

—NASA/JPL

MARTIAN HISTORICAL TIME

Scholars have identified historical ages in various cultures on Earth; for instance, in Europe, there are Antiquity (ancient Greece and Rome), the Middle Ages, the Renaissance, the Age of Discovery, the Enlightenment, the Industrial Revolution, and so on. It is also possible to construct a time scale for Mars. In his 1892 book, *La planete Mars et ses conditions d'habitabilite,* the French astronomer Camille Flammarion identified three historical periods: 1636-1829, 1830-1876, and 1877-1892:

> The first period of what we may call the history of Mars began with the first telescopic views of the planet, obtained in the seventeenth century.
>
> The second period of our study begins with the great areographic work of Beer and Mädler, the first continuous observations that allowed the authors to draw up a geographical chart of Mars.
>
> Our third period includes the Martian cycle from 1877 to 1892, which has been more fruitful than any of those preceding it.

I expand on Flammarion's schema, as there was some Martian history before 1636, and there certainly has been a lot of Martian history since 1892. Although there may be no persuasive argument for expressing dates in Martian terms prior to 1609, the long history of observing Mars with the naked eye deserves recognition and study. This Gymnoptic Era can further be divided into three periods. The first of these is the Mythic Period, when Mars was perceived as one of the many supernatural apparitions in the sky. This is followed by the Astrologic Period, when observers recorded the motions of Mars and the other planets in the hope of predicting Earthly events. The Cosmologic Period, when observers recorded the motions of Mars and the other planets in the quest to understand the order of the cosmos, closes out the Gymnoptic Era. The Darian calendar begins with the Telescopic Era, which can be divided into two periods. The Monographic Period, when written descriptions of Mars dominated the observation records, together with some drawings of Martian albedo features, encompasses Flammarion's first period. The Telescopic Era ends with the Cartographic Period, when astronomers' maps and photographs of Mars were the dominant mode of recording observations of Mars, comprises Flammarion's second and third periods, and it might be useful to define later subdivisions of this period. Telescopic observations of Mars certainly did not end with *Mariner 4's* flyby of Mars in 1965, but from that time forward the telemetry from our automated spacecraft have been the primary means of obtaining scientific data about Mars. The Telemetric Era begins with the Reconnaissance Period, during which telemetric data returned from Mars missions was intermittent, followed by the present the present Surveillance Period, which is characterized by overlapping missions which have provided a continuous stream of data from Mars. We can look forward to a future Anthropic Era, when humans will first explore the surface of Mars and return to Earth (the Expeditionary Period), establish colonies with permanent populations under the jurisdiction of either individual terrestrial nation-states or collective entities thereof (the Colonial Period), and eventually form sovereign, self-governing states (the Independence Period). The Darian calendar is intended to serve as a schema for studying history in a

Martian context and chronicling the present activities of robotic explorers on the surface of Mars, as well as to provide for the timekeeping needs of an emerging Martian culture.

Era		Period		Gregorian Date	Darian Date
Gymnoptic	Mars as studied in naked eye astronomy	Mythic	Mars as perceived by prehistoric cultures	2500th - 30th c. BCE	undefined
		Astrologic	Observation of Mars to predict Earthly events	29th - 7th c. BCE	
		Cosmologic	Observation of Mars to understand the order of the cosmos	6th c. BCE - 1608	
Telescopic	Mars as observed through telescopes	Monographic	Written descriptions and drawings of Mars's appearance	1609 - 1829	0 - 116
		Cartographic	Maps and Earth-based photography of the Martian surface	1830 - 1964	117 - 188
Telemetric	Mars as investigated via robotic spacecraft	Reconnaissance	Sporadic telemetric investigation of Mars	1965 - 1996	189 - 205
		Surveillance	Continuous telemetric coverage of Mars	1997 -?	206 -?
Anthropic	Mars as explored and settled by humans	Expeditionary	Human expeditions to and bases on Mars	?	?
		Colonial	Permanent population on Mars subject to Earth jurisdiction	?	?
		Independence	Sovereign self-government on Mars	?	?

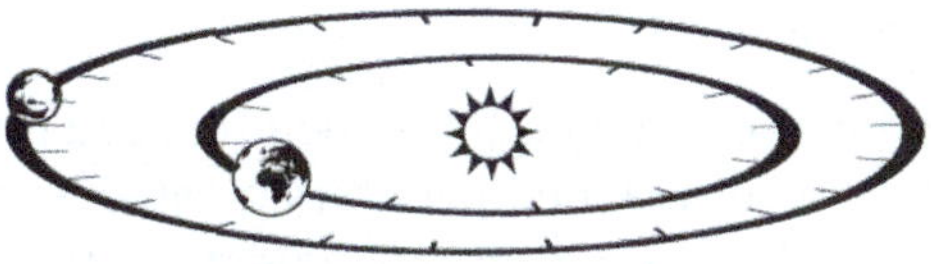

221 PISCES

As seen from Mars, the Sun slowly traverses the western sector of the broad constellation Pisces as Mars has recently passed aphelion. In Western tradition, Pisces is depicted as two fishes tied together on a knotted cord.

The median one-way communications delay during this month is 13 minutes 1 second.

221 Pisces 01 **2025 May 02 10:46 UTC**

Ls 078.0° Curiosity Sol 4528
MSD 52797 Perseverance Sol 1493

The first sol of the first quarter of Summer on the Aitken calendar; approximately the first sol of December on the Heinlein calendar.

On this sol in 210 (2004 August 22):

MER-B Sol 207: The *Opportunity* team entered restricted planning. This happens when the timing of the rover's sol on Mars and the day in the California time zone get out of sync due to the nearly 40-minute difference in length of Earth days and Mars sols. The afternoon downlink arrives at JPL too late in the day to plan the next sol unless the team works through the night. Instead of staying up all night, the team plans with restrictions that forbid rover movement or arm activity on a sol immediately following a sol on which the rover has moved. This gives additional time for the data to become available so that planners can use up-to-date knowledge about the rover's position and orientation.

So, rather than any driving on Sol 207, *Opportunity* conducted remote-sensing work, including atmospheric observations and panoramic camera imaging of several features.

On this sol in 210 (2004 August 23):

MER-A Sol 228: *Spirit* continued its examination of Ebenezer.

—NASA/JPL

221 Pisces 02 **2025 May 03 11:26 UTC**

Ls 078.4° Curiosity Sol 4529
MSD 52798 Perseverance Sol 1494

Kalendis Junius on the Roman calendar (analogous to June 01 on the recalibrated Pickering calendar) was sacred to Mars. It was the anniversary of the dedication of the Templum Martis (Temple of Mars) near the Capena gate in Rome.

On this sol in 153 (1897 June 09):

"Whitish at pole."

—Andrew E. Douglass

On this sol in 210 (2004 August 23):

MER-B Sol 208: *Opportunity* drove again. It bumped forward to put Escher well within the arm's work volume. The sol also included panoramic camera imaging of Escher and of a trench created by *Opportunity's* prior wheel movements in the vicinity. *Opportunity* slept deeply on the night of Sol 208 for the second night in a row. The purpose of successive deep sleeps was to align the deep-sleep nights with poorer overnight *Mars Odyssey* passes, leaving the rover ready to take advantage of higher-volume passes on alternate nights.

On this sol in 210 (2004 August 24):

MER-A Sol 229: *Spirit* continued its examination of Ebenezer.

—NASA/JPL

221 Pisces 03 **2025 May 04 12:05 UTC**

L_S 078.9° Curiosity Sol 4530
MSD 52799 Perseverance Sol 1495

On this sol in 183 (1953 November 14):

Rod Brown of the Rocket Rangers Season 1 Episode 31 "The Magic Man of Mars" airdate. A carnival entertainer who specializes in magic has a more lucrative secret life as a space pirate.

—International Movie Database

On this sol in 210 (2004 August 25):

MER-A Sol 209: Activities for *Opportunity* focused on recharging its batteries and downlinking data through both an afternoon communications link with the *Mars Odyssey* orbiter and an overnight *Mars Odyssey* pass (early morning Sol 210). The rover also made some remote sensing observations. The Sol 209 plan was constructed without benefit of the Sol 208 downlink (due to the restricted nature of the planning at this point in the cycle of Martian sols progressing relative to Earth days). For that reason, activities were limited to remote sensing, which did not depend upon knowledge of the exact position of the rover. All went fine, and about 165 megabits of data were returned through relay by *Mars Odyssey*.

MER-A Sol 230: *Spirit* used the rock abrasion tool to brush a target on Ebenezer and took an overnight alpha particle X-ray spectrometer reading on the brushed area.

—NASA/JPL

221 Pisces 04 **2025 May 05 12:45 UTC**

L_S 079.3° Curiosity Sol 4531
MSD 52800 Perseverance Sol 1496

This sol is sacred to Bellona (sister of Mars) on the Roman calendar, celebrated on ante dies III Nonas Junius (analogous to June 03 on the recalibrated Pickering calendar).

On this sol in 146 (1884 April 12):

Alfred Johanne "Bob" Olsen was born; American science fiction writer; author of "The Space Marines and the Slavers" (1936).

On this sol in 184 (1955 October 03):

Journey into Space: The World in Peril, Episode 2 airdate. Jet and Lemmy go to investigate the objects in an orbital rocket, but are sent to sleep for 24 hours by a strange noise. Film footage suggests the asteroids are actually spacecraft carriers. Jet discovers that a spaceship has crashed in the Lake District. Jet and Doc enter the ship.

—Wikipedia

On this sol in 210 (2004 August 26):

MER-B Sol 210: *Opportunity* was commanded to image the rock abrasion tool in a variety of positions to get better knowledge of a pebble that appeared to be jammed between rotors in an earlier image. Then the rover used its microscopic imager to survey several spots on a rock called Escher. In addition, it took panoramic camera images for assembling into a mosaic of the dune field at the bottom of Endurance Crater. The diagnostic images of the rock abrasion tool brought good news with the revelation that there is no longer a pebble jammed between the grind bits! Apparently, sometime after the last previous images of the tool were acquired on Sol 200 (210 Kumbha 21), the pebble fell out, perhaps due to thermal cycling or vehicle motion. The team will proceed next with diagnostic actions to confirm that the abrasion tool is functioning normally again. Those activities are planned for Sol 212.

MER-A Sol 231: *Spirit's* abrasion tool ground on Ebenezer for two hours into the same spot it had brushed.

—NASA/JPL

221 Pisces 05 | **2025 May 06 13:24 UTC**

Ls 079.8° — Curiosity Sol 4532

MSD 52801 — Perseverance Sol 1497

On this sol in 152 (1895 July 27):

James Milton Wilcox (Jayem Wilcox) was born; American pulp artist; created the interior art for "Shambleau" by C.L. Moore (1933).

The Boston Daily Globe

RADIO PLAY TERRIFIES NATION

On this sol in 175 (1938 October 30):

The Mercury Theater on the Air, starring Orson Welles, performed Howard Koch's radio play "Invasion from Mars," based on H.G. Wells's *The War of the Worlds.* The broadcast terrified the USA and Canada.

On this sol in 210 (2004 August 27):

MER-B Sol 211: *Opportunity* awoke from deep sleep at 7 a.m. local solar time. It re-enabled survival heaters on its miniature thermal emission spectrometer and re-started a Mössbauer spectrometer examination of a target called Kirchner. The rover made observations with its panoramic camera and its miniature thermal emission spectrometer from about 11 a.m. to 2 p.m. local solar time, focusing on getting thermal inertia measurement of the dunes at different times of day. A planned tool change to the alpha particle X-ray spectrometer in the afternoon failed due to a sequencing error in retracting the Mössbauer spectrometer from the surface. A conditional sequencing check prevented the overnight alpha particle X-ray spectrometer integration from occurring as desired in such fault cases.

MER-A Sol 232: *Spirit* completed an overnight reading with its alpha particle X-ray spectrometer on a target dubbed Cratchit 2, where the rover had earlier used its rock abrasion tool to cut a hole exposing the rock's interior. *Spirit* took images of the same target with the microscopic imager. *Spirit* then placed the Mössbauer spectrometer on Cratchit 2 and started a very long, two-sol integration.

—NASA/JPL

221 Pisces 06 **2025 May 07 14:04 UTC**

Ls 080.2° Curiosity Sol 4533

MSD 52802 Perseverance Sol 1498

On this sol in 175 (1938 October 31):

RADIO LISTENERS IN PANIC, TAKING WAR DRAMA AS FACT

Message From the Police

The New York police sent out the following:

"To all receivers: Station WABC informs us that the broadcast just concluded over that station was a dramatization of a play. No cause for alarm."

The New Jersey State Police teletyped the following:

"Note to all receivers—WABC broadcast as drama re this section being attacked by residents of Mars. Imaginary affair."

From one New York theatre a manager reported that a throng of playgoers had rushed from his theatre as a result of the broadcast. He said that the wives of two men in the audience, having heard the broadcast, called the theatre and insisted that their husbands be paged. This spread the "news" to others in the audience.

The switchboard of *The New York Times* was overwhelmed by the calls. A total of 875 were received. One man who called from Dayton, Ohio, asked, "What time will it be the end of the world?" A caller from the suburbs said he had had a houseful of guests and all had rushed out to the yard for safety.

Warren Dean, a member of the American Legion living in Manhattan, who telephoned to verify the "reports," expressed indignation which was typical of that of many callers.

"I've heard a lot of radio programs, but I've never heard anything as rotten as that," Mr. Dean said. "It was too realistic for comfort. They broke into a dance program with a news flash. Everybody in my house was agitated by the news. It went on just like press radio news."

At 9 o'clock a woman walked into the West Forty-seventh Street police station dragging two children, all carrying extra clothing. She said she was ready to leave the city. Police persuaded her to stay.

A garbled version of the reports reached the Dixie Bus Terminal, causing officials there to prepare to change their schedule on confirmation of "news" of an accident at Princeton on their New Jersey route. Miss Dorothy Brown at the terminal sought verification, however, when the caller refused to talk with the dispatcher, explaining to her that "the world is coming to an end and I have a lot to do."

—*The New York Times,* page 1

On this sol in 187 (1961 May 26):

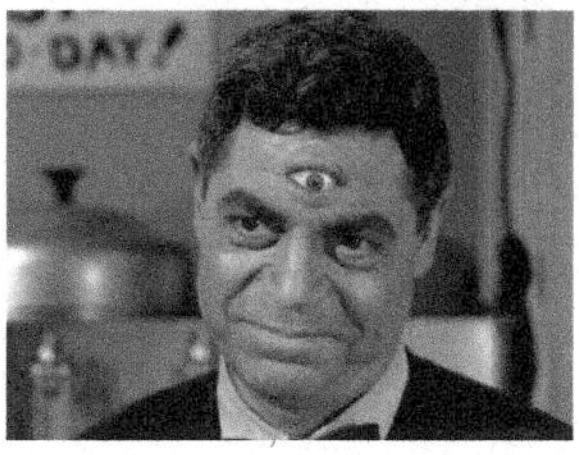

The Twilight Zone, Season 2 Episode 28, "Will the Real Martian Please Stand Up?" air date. Wintry February night, the present. Order of events: a phone call from a frightened woman notating the arrival of an unidentified flying object, then the checkout you've just witnessed, with two state troopers verifying the event—but with nothing more enlightening to add beyond evidence of some tracks leading across the highway to a diner. You've heard of trying to find a needle in a haystack? Well, stay with us now, and you'll be part of an investigating team whose mission is not to find that proverbial needle, no, their task is even harder. They've got to find a Martian in a diner, and in just a moment you'll search with them, because you've just landed—in *The Twilight Zone.*

—Rod Serling

On this sol in 210 (2004 August 28):

MER-B Sol 212: Engineers successfully completed a calibration of *Opportunity's* rock abrasion tool. The tool is healthy and ready for action. An aggressive plan acquired 80 microscopic images of the rock Escher. As part of the team's efforts to increase operational flexibility, a test was conducted involving operating the miniature thermal emission spectrometer in parallel with arm operations. Unfortunately, this resulted in some corrupted data from the miniature thermal emission spectrometer due to vibrations as the rover arm moved. The rover used its Mössbauer spectrometer in the afternoon before going into deep sleep overnight.

—NASA/JPL

221 Pisces 07 **2025 May 08 14:44 UTC**

Ls 080.7° Curiosity Sol 4534
MSD 52803 Perseverance Sol 1499

On this sol in 151 (1893 September 10):

Aleksei Mikhailovich Fajko (Алексей Михайлович Файко) was born; Russian/Soviet screenwriter; contributing writer to the motion picture *Aelita: Queen of Mars (Аэлита – королева Марса)* (1924).

On this sol in 210 (2004 August 29):

MER-B Sol 213: *Opportunity* awoke from deep sleep and re-started the Mössbauer integration. The rover performed some remote sensing during the day and then changed tools to the alpha particle X-ray spectrometer for an overnight integration. Later it completed a midnight thermal inertia observation with the miniature thermal emission spectrometer, which required an hour of actuator preheating.

MER-A Sol 234: After the two-sol Mössbauer spectrometer activity, *Spirit* spent the sol performing 90 minutes of remote sensing.

—NASA/JPL

221 Pisces 08 **2025 May 09 15:23 UTC**

Ls 081.1° Curiosity Sol 4535
MSD 52804 Perseverance Sol 1500

On this sol in 139 (1871 February 14):

In *Thuvia, Maid of Mars* (1916) by Edgar Rice Burroughs, Thuvia of Ptarth was enslaved by the therns. According to Ekman's chronology, the Barsoomian date was 2 ord, 422 padan.

On this sol in 210 (2004 August 30):

MER-B Sol 214: *Opportunity* completed the alpha particle X-ray spectrometer observation and successfully used the rock abrasion tool to brush clean two targets on Escher (EmilNolde and Kirchner_RAT). Then it made observations with its microscopic imager, hazard-avoidance camera and panoramic camera. The Mössbauer spectrometer was then positioned on Kirchner_RAT, where it analyzed the rock's mineral composition until the rover went into deep sleep overnight.

MER-A Sol 235: *Spirit* changed tools from the Mössbauer spectrometer to the microscopic imager and took pictures. Then *Spirit* used the alpha particle X-ray spectrometer for thirty minutes. After completing the alpha particle X-ray spectrometer reading, the arm was stowed so that on Sol 236 (210 Pisces 09) the rock abrasion tool brushings could be imaged using the panoramic camera without the arm blocking any part of the image.

—NASA/JPL

221 Pisces 09 **2025 May 10 16:03 UTC**

Ls 081.5° Curiosity Sol 4536
MSD 52805 Perseverance Sol 1501

The first sol of July, marsrev 3679 on the Levitt calendar.

On this sol in 210 (2004 August 31):

MER-B Sol 215: *Opportunity* completed a reading with its Mössbauer spectrometer of a target called Kirchner, where a wire brush on the rover's rock abrasion tool had scrubbed a circular patch on the surface of a rock called Escher. The rover also made some remote-sensing observations then then set up for using its alpha particle X-ray spectrometer on Kirchner early the following morning. However, an image from the rover's hazard-avoidance camera revealed that the doors of the alpha particle X-ray spectrometer had not completely opened. The door is a tricky mechanism; incomplete openings and closings have occurred before, and the team continues to work on approaches to more reliably maneuver the door.

MER-A Sol 236: *Spirit* spent the sol brushing eight adjoining patches on Ebenezer to create a large enough scrubbed area for analyzing with the miniature thermal emission spectrometer.

—NASA/JPL

221 Pisces 10 **2025 May 11 16:42 UTC**
Ls 082.0° Curiosity Sol 4537
MSD 52806 Perseverance Sol 1502

On this sol in 141 (1874 November 22):

Harry Gore Bishop was born; American artillery general and writer; author of "On the Martian Way" (1907). It is sometime in the future when space runs to the other planets are routine matters, together with a flourishing interplanetary economy. The R.D. Jones space line has been having difficulties, what with soft business and the loss of a ship to a meteor swarm. The company therefore decides to convert the *Columbia* into a passenger ship, which is to leave Nekhoboh (Hoboken) with invalids who are going to Mars for the benefits of the rarefied air.

—Everett Franklin Bleiler, 1990

On this sol in 146 (1884 April 18):

Rudolph Franz Paul (Frank R. Paul) was born; Hungarian-American artist; was influential in defining the look of both cover art and interior illustrations in the science fiction pulps of the 1920s and 1930s; inducted into the Science Fiction Hall of Fame in 2009.

On this sol in 152 (1895 August 01):

Part 4 of *Mars* by Percival Lowell appeared in *The Atlantic* (date approximate).

On this sol in 210 (2004 September 01):

MER-B Sol 216: *Opportunity* successfully acquired early morning alpha particle X-ray spectrometer data on Kirchner. Despite the incomplete opening of the instrument's dust doors, the spectra look good. No repeat of the integration will be necessary. The rover also used the Mössbauer spectrometer to examine another brushed target, EmilNolde, on Escher. This reading was planned to run into the evening then later, following a deep sleep, to resume in the early morning of Sol 217. The Mössbauer placement went fine. The rover was commanded to close and reopen the alpha particle X-ray spectrometer doors and this went well. The doors are now properly open and ready for action on Sol 217.

MER-A Sol 237: *Spirit* bumped backwards and took some navigation camera images of the brushed area of Ebenezer to support miniature thermal emission spectrometer observations on the next sol. A navigation camera panorama was taken in the expected drive direction, toward Tikal.

—NASA/JPL

221 Pisces 11 **2025 May 12 17:22 UTC**

Ls 082.4° Curiosity Sol 4538

MSD 52807 Perseverance Sol 1503

Mars arrival opportunities from Earth via a Type I trajectory and an EEM trajectory.

Earth arrival opportunity from Mars via a Type II trajectory.

On this sol in 153 (1897 June 19):

Moses Horwitz (Moe Howard) was born; American actor, screenwriter, and producer; starred in the motion pictures *Space Master X-7* (1958) and *The Three Stooges in Orbit* (1962).

On this sol in 155 (1901 March 24):

Ub Iwerks was born; American animator; directed the animated short film "Scrappy's Trip to Mars" (1937).

On this sol in 184 (1955 October 10):

Journey into Space: The World in Peril, Episode 3 airdate. The crew and the Lunar Controller try to locate the crew of the crashed ship. They interview Moore, a commercial traveller who may have had contact with the crew of the ship, who recounts his experience. The crew are ordered back to Mars and are driven to an airfield, but the car crashes when the crew tackle the "Conditioned Type" driver.

—Wikipedia

On this sol in 210 (2004 September 02):

MER-B Sol 217: *Opportunity* used its rock abrasion tool to brush a target called Otto Dix, and used its microscopic imager to look at the brushed area. Then the rover was commanded to place the alpha particle X-ray spectrometer on EmilNolde, precisely on a dirty portion of that target (an area that was not very well cleared away by the brush action a few sols ago). The plan was to collect data with the alpha particle X-ray spectrometer in the evening, perform a move in the middle of the night to a cleanly brushed portion of EmilNolde and integrate again until morning. These two integrations will be used to discern the differences between the clean and dirty portions of the target. A 100-megabit afternoon downlink through *Mars Odyssey* on Sol 217 showed that all activities went well through the placement of the alpha particle X-ray spectrometer on the dirty part of EmilNolde.

MER-A Sol 238: The panoramic camera imaged the brushed area of Ebenezer. A little after noon, the miniature thermal emission spectrometer took spectral readings of both the brushed area and a hole that rock abrasion tool hole had cut into Ebenezer on Sol 231 (210 Pisces 04). At 2 p.m. Gusev time, *Spirit* took an afternoon drive to Tikal, about nine meters (30 feet) away.

Solar conjunction minus 11. *Spirit* will spend solar conjunction at Tikal. Solar conjunction is when Earth and Mars are on opposite sides of the Sun. Due the interference with the Sun, communications between Earth and the spacecraft at Mars will be minimal during solar conjunction, which occurs during *Spirit's* Sols 244 through 255 (210 Pisces 17 through 28).

—NASA/JPL

221 Pisces 12 **2025 May 13 18:01 UTC**

Ls 082.9° Curiosity Sol 4539

MSD 52808 Perseverance Sol 1504

On this sol in 149 (1889 December 10):

Ray Bidwell Collins was born; American actor; starred in The Mercury Theatre on the Air's radio play "Invasion from Mars" (1938).

On this sol in 173 (1935 February 01):

"The Robot Aliens" by Earl Binder and Otto Binder (as Eando Binder) appeared in *Wonder Stories* (date approximate). The Earth is invaded by Martian robots who hope to succeed where H.G. Wells's Martians failed....

Cover and interior art by Frank R. Paul.

Part 4 of *Swords of Mars* by Edgar Rice Burroughs appeared in *The Blue Book Magazine* (date approximate).

Interior art by Robert Fink.

221 Pisces 13 **2025 May 14 18:41 UTC**

Ls 083.3° Curiosity Sol 4540
MSD 52809 Perseverance Sol 1505

On this sol in 175 (1938 November 7):

The serial motion picture *Flash Gordon's Trip to Mars* (also known as *Mars Attacks the World)* was released. When a deadly Nitron ray strikes Earth, Flash Gordon and his friends travel to Mars to battle Ming the Merciless and his new ally Queen Azura.

—International Movie Database

221 Pisces 14 **2025 May 15 19:21 UTC**

Ls 083.8° Curiosity Sol 4541
MSD 52810 Perseverance Sol 1506

On this sol in 210 (2004 September 05):

MER-B Sol 220: This overview of Endurance Crater traces the path of the Mars Exploration Rover *Opportunity* from Sol 94 (210 Dhanus 27) to Sol 220.

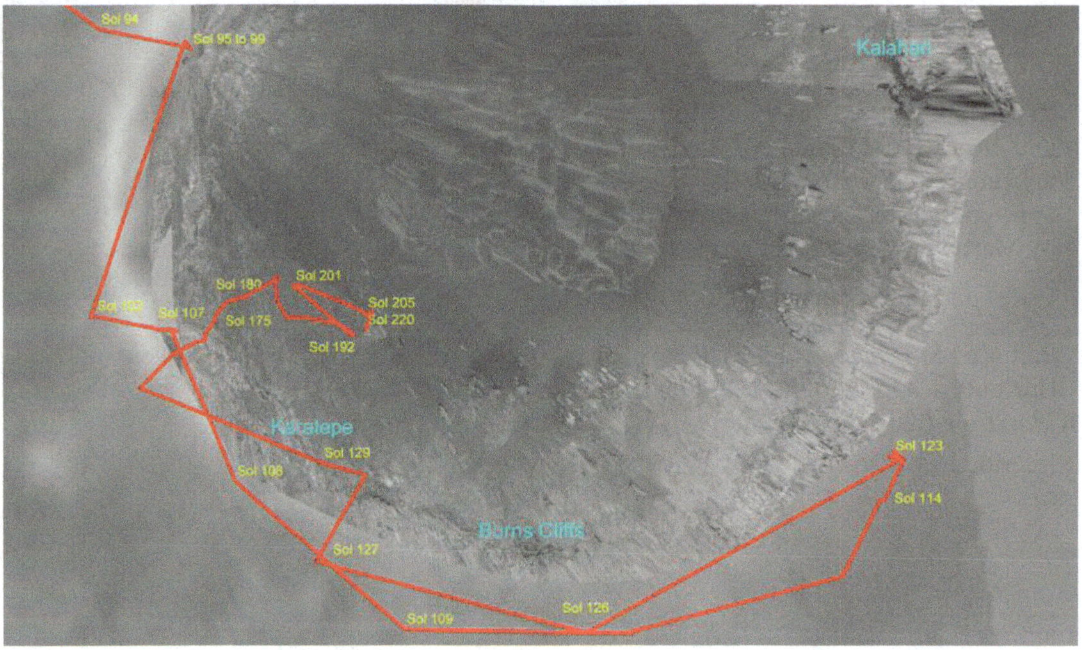

MER-A Sol 241: Solar conjunction minus 8. *Spirit* is in safe place to continue daily science observations automatically throughout the solar conjunction period when engineers and scientists will be unable to send commands reliably to the rover. An 18-sol period began a transition into solar conjunction on Sol 241, when the Sun partially obscured the communications path between Earth and Mars, making communications sessions unreliable. Engineers were able to successfully command *Spirit* on Sol 241.

—NASA/JPL

221 Pisces 15 **2025 May 16 20:00 UTC**

L_s 084.2° Curiosity Sol 4542
MSD 52811 Perseverance Sol 1507

On this sol in 176 (1940 September 27):

Edgar Rice Burroughs began writing "The Black Pirates of Barsoom," the second part of a new Mars series.

On this sol in 210 (2004 September 06):

MER-A Sol 242: Solar conjunction minus 7. Engineers had partial commanding success with *Spirit* prior to solar conjunction.

Engineers sent *Spirit* a set of coordinated commands to use the miniature thermal emission spectrometer and panoramic camera for observations of possible future science targets. A new set of 12 conjunction master sequences was also transmitted successfully to *Spirit*. This new set of conjunction master sequences will use less energy than previous sequences.

Since *Spirit* arrived at its solar-conjunction resting place, its science activities have focused on gathering data from the surrounding area for use in planning post-conjunction sols. Navigation camera images in *Spirit's* drive direction have been used to develop traverse maps. These maps show areas that allow *Spirit* to maintain a north-facing tilt; these areas will provide significantly more solar energy and will therefore be favored as the team plans the traverse to *Spirit's* next science target.

Engineers and scientists decided to leave the Mössbauer in place on the magnet and will evaluate the status and effect of the dirt clod after conjunction. The dirt does not pose any threat to the rover from an engineering perspective.

For the conjunction period, the rover team has placed the Mössbauer spectrometer on one of the two magnets on the rover deck. *Spirit* will activate the Mössbauer instrument every day during conjunction in order to characterize the dust that has collected on the magnet. However, a wrinkle has developed in this plan. Before placing the Mössbauer spectrometer on the magnet, *Spirit* placed it on the soil in front of its current location. That soil touch was done to leave a soil impression that would be studied after conjunction for changes. Images taken after the touch indicate that *Spirit* inadvertently picked up some soil and likely sandwiched that soil onto the magnet with the Mössbauer. It's the team's first inadvertent sample acquisition.

During conjunction, *Spirit* will transmit five-minute "beep" tones, and engineers will send "No-operation" commands to the rover to characterize effects that the conjunction has on radio transmissions between Mars and Earth.

—NASA/JPL

221 Pisces 16 **2025 May 17 20:40 UTC**

L_s 084.7° Curiosity Sol 4543
MSD 52812 Perseverance Sol 1508

Possible meteor shower associated with parent body C/1854 L1 Klinkerfues: Martian right ascension 245.1°, declination 39.8°.

On this sol in 210 (2004 September 07):

MER-A Sol 243: This was the last sol of normal commanding for *Spirit* before conjunction. The rover team successfully re-transmitted four conjunction sequences that had not made it on-board during the Sol 242 uplink. The team saw no transmission errors (but commanded everything twice just in case), and the rover performed the commanded remote sensing science.

—NASA/JPL

221 Pisces 17 **2025 May 18 21:19 UTC**
Ls 085.2° Curiosity Sol 4544
MSD 52813 Perseverance Sol 1509

On this sol in 184 (1955 October 17):

Journey into Space: The World in Peril, Episode 4 airdate. The crew fly to Australia, then on a ship to the Moon. While preparing for the voyage to Mars, they spot unidentified objects on the lunar horizon. The takeoff is complicated by strange communication with control, suggesting the Moon base has been infiltrated.

—Wikipedia

On this sol in 210 (2004 September 08):

MER-B Sol 223: Solar conjunction minus 6. Command sequences previously sent to *Opportunity* performed a set of automatic science activities while communications with Earth were disrupted during the solar conjunction period.

MER-A Sol 244. The rover team transitioned *Spirit* into conjunction operations and did the first "no-op" commanding tests during midsol to see how effective the command link was as Mars moved further behind the Sun. The team received data from the *Mars Odyssey* orbiter indicating that *Spirit* was healthy and proceeding normally with on-board conjunction sequences.

—NASA/JPL

221 Pisces 18 **2025 May 19 21:59 UTC**
Ls 085.6° Curiosity Sol 4545
MSD 52814 Perseverance Sol 1510

On this sol in 210 (2004 September 09):

MER-B Sol 224: Solar conjunction minus 5. *Opportunity* team transitioned into conjunction operations and did the first "no-op" commanding tests during midday to see how effective the command link was as Mars moved further behind the Sun. The team received data from the *Mars Odyssey* orbiter indicating that *Opportunity* was healthy and proceeding normally with on-board conjunction sequences.

MER-A Sol 245: No commanding was done. *Spirit* automatically took daily atmospheric science measurements and made filter magnet observations with the Mössbauer spectrometer.

—NASA/JPL

221 Pisces 19 | **2025 May 20 22:39 UTC**

Ls 086.1° | Curiosity Sol 4546
MSD 52815 | Perseverance Sol 1511

On this sol in 176 (1940 October 01):

"Oscar, Detective of Mars" by James Norman Schmidt (as James Norman) appeared in *Fantastic Adventures* (date approximate). Out of a magician's hat popped the lovable Martian, Oscar, right smack into the middle of a deadly situation.

Interior artwork uncredited.

On this sol in 210 (2004 September 10):

MER-B Sol 225, MER-A Sol 246: Solar conjunction minus 4. Command sequences previously sent to *Opportunity* and *Spirit* performed a set of automatic science activities while communications with Earth were disrupted during the solar conjunction period.

—NASA/JPL

221 Pisces 20 | **2025 May 21 23:18 UTC**

Ls 086.5° | Curiosity Sol 4547
MSD 52816 | Perseverance Sol 1512

Predicted meteor shower associated with parent body C/2007 D2 Spacewatch: Martian right ascension 236.3°, declination 7.4°. Impacting at 53 km/sec on the morning side.

On this sol in 210 (2004 September 11):

MER-B Sol 226, MER-A Sol 247: Solar conjunction minus 3. Command sequences previously sent to *Opportunity* and *Spirit* performed a set of automatic science activities while communications with Earth were disrupted during the solar conjunction period.

—NASA/JPL

221 Pisces 21 | **2025 May 22 23:58 UTC**

Ls 087.0° | Curiosity Sol 4548
MSD 52817 | Perseverance Sol 1513

On this sol in 210 (2004 September 12):

MER-B Sol 227

On this sol in 210 (2008 September 13):

MER-A Sol 248

Solar conjunction minus 2. Command sequences previously sent to *Opportunity* and *Spirit* performed a set of automatic science activities while communications with Earth were disrupted during the solar conjunction period.

—NASA/JPL

221 Pisces 22 **2025 May 24 00:37 UTC**
Ls 087.4° Curiosity Sol 4549
MSD 52818 Perseverance Sol 1514

On this sol in 144 (1880 July 27):

Austin Javen Hall was born; American science fiction writer; author of “The Man Who Saved the Earth” (1919).

On this sol in 210 (2004 September 13):

MER-B Sol 228

On this sol in 210 (2008 September 14):

MER-A Sol 249

Solar conjunction minus 1. Command sequences previously sent to *Opportunity* and *Spirit* performed a set of automatic science activities while communications with Earth were disrupted during the solar conjunction period.

—NASA/JPL

221 Pisces 23 **2025 May 25 01:17 UTC**
Ls 087.9° Curiosity Sol 4550
MSD 52819 Perseverance Sol 1515

On this sol in 153 (1897 July 01):

Part 10 (the conclusion) of *The Martian* by George du Maurier appeared in *Harper’s New Monthly Magazine* (date approximate).

On this sol in 205 (1995 April 21):

The Monuments of Mars by Mark Dwane was reissued on CD. Set list:

1 Eternity
2 Pyramids of Mars
3 Metamorphosis
4 Solstice Ritual
5 Cydonia
6 Water Age
7 Pyramids of Mars, Reprise
8 The Martian Sphinx
9 Another Eternity

On this sol in 210 (2004 September 14):

MER-B Sol 229

On this sol in 210 (2008 September 15):

MER-A Sol 250

Solar conjunction occurred at 15:57 AMT (12:55 UTC). Command sequences previously sent to *Opportunity* and *Spirit* performed a set of automatic science activities while communications with Earth were disrupted during the solar conjunction period.

—NASA/JPL

221 Pisces 24 **2025 May 26 01:57 UTC**
Ls 088.3° Curiosity Sol 4551
MSD 52820 Perseverance Sol 1516

On this sol in 184 (1955 October 24):

Journey into Space: The World in Peril, Episode 5 airdate. *Discovery* gets away safely; however, they cannot establish contact with Moon control, so Lemmy contacts Earth. After ships from Earth to the Moon are lost, the crew concludes that the Moon has fallen under Martian control. While refuelling from freighter #1, Jet and Lemmy hear a strange wailing voice.

—Wikipedia

On this sol in 185 (1957 September 09):

Studio One in Hollywood episode "The Night America Trembled" airdate. On October 30, 1938, Orson Welles's Mercury Players performed a radio play version of H.G. Wells's *The War of the Worlds*. Some people in the unsuspecting public took the broadcast as a real report about a Martian invasion. In this television play of twenty years later, several fictional storylines present the story of the 1938 broadcast and the panic it caused. Journalist Edward R. Murrow, as himself, provides short narrative bridges between scenes. The fun includes watching the young extras for a surprising number of faces that will later range from the familiar to the famous.

—International Movie Database

On this sol in 210 (2004 September 15):

MER-B Sol 230

On this sol in 210 (2008 September 16):

MER-A Sol 251

Solar conjunction plus 1. Command sequences previously sent to *Opportunity* and *Spirit* performed a set of automatic science activities while communications with Earth were disrupted during the solar conjunction period.

—NASA/JPL

221 Pisces 25 **2025 May 27 02:36 UTC**

Ls 088.8° Curiosity Sol 4552
MSD 52821 Perseverance Sol 1517

On this sol in 210 (2004 September 16):

MER-B Sol 231

On this sol in 210 (2008 September 17):

MER-A Sol 252

Solar conjunction plus 2. Command sequences previously sent to *Opportunity* and *Spirit* performed a set of automatic science activities while communications with Earth were disrupted during the solar conjunction period.

—NASA/JPL

221 Pisces 26 **2025 May 28 03:16 UTC**

Ls 089.2° Curiosity Sol 4553
MSD 52822 Perseverance Sol 1518

Mars arrival opportunities from Earth via a Type I trajectory and an EEM trajectory.

On this sol in 184 (1955 October 26):

Shorty Rogers and His Giants recorded the songs "Martians Come Back" and "March of the Martians."

On this sol in 210 (2004 September 17):

MER-B Sol 232

On this sol in 210 (2008 September 18):

MER-A Sol 253

Solar conjunction plus 3. Command sequences previously sent to *Opportunity* and *Spirit* performed a set of automatic science activities while communications with Earth were disrupted during the solar conjunction period.

—NASA/JPL

221 Pisces 27 **2025 May 29 03:55 UTC**

Ls 089.7° Curiosity Sol 4554
MSD 52823 Perseverance Sol 1519

Northern solstice occurs at 18:48 AMT (23:15 UTC).

Quinquatrus Minores, a festival sacred to Minerva and Mars, celebrated on ante dies pridie Idus Junius on the Roman calendar (analogous to June 26 on the recalibrated Pickering calendar).

On this sol in 171 (1931 May 15):

A Fighting Man of Mars by Edgar Rice Burroughs was published as a novel. Hadron of Hastor, Padwar in the service of the Warlord of Mars, goes alone in search of the beautiful Sanoma Tora when she is abducted by Tul Axtar of Jahar. But before Hadron manages to reach Jahar he is deflected from his purpose in an effort to rescue a slave girl of his own race from the ferocious green men of Mars. Tavia, for that is the lovely girl's name, leads Hadron to her home city Tjanath from which she had been stolen as a very small girl.

—Edgarriceburroughs.com

Cover art by Hugh Hutton.

On this sol in 172 (1933 April 01):

"The Forgotten Man of Space" by P. Schuyler Miller appeared in *Wonder Stories* (date approximate).

Interior art by Marco Marchioni.

On this sol in 201 (1987 October 16):

The song "Couple of Mates (From Mars & Jupiter)" by the British Manfred Mann's Earth Band was released on the CD *Masque.* The song included sketches of the "Mars" and "Jupiter" movements from Gustav Holst's *The Planets.*

On this sol in 210 (2004 September 18):

MER-B Sol 233

On this sol in 210 (2008 September 19):

MER-A Sol 254

Solar conjunction plus 4. Command sequences previously sent to *Opportunity* and *Spirit* performed a set of automatic science activities while communications with Earth were disrupted during the solar conjunction period.

—NASA/JPL

221 Pisces 28 **2025 May 30 04:35 UTC**

Ls 090.1° Curiosity Sol 4555

MSD 52824 Perseverance Sol 1520

On this sol in 179 (1946 June 01):

"The Million Year Picnic" by Ray Bradbury appeared in Planet Stories (date approximate). They were supposed to be starting on a picnic, a wonderful day of fun. But there was a gun in the boat, and too much food and equipment. And just behind the veil of vacation—instead of the soft face of laughter—there was something hard and bony and terrifying.

Later incorporated into *The Martian Chronicles* as the final vignette, the story ends famously:

> They reached the canal. It was long and straight and cool and wet and reflective in the night.
>
> "I've always wanted to see a Martian," said Michael. "Where are they, Dad? You promised."
>
> "There they are," said Dad, and he shifted Michael on his shoulder and pointed straight down.
>
> The Martians were there. Timothy began to shiver.
>
> The Martians were there—in the canal—reflected in the water. Timothy and Michael and Robert and Mom and Dad.
>
> The Martians stared back up at them for a long, long silent time from the rippling water....

On this sol in 188 (1963 May 06):

NASA deferred Mariner B, an Atlas-Centaur launch of an unmanned Mars flyby mission, formerly slated for 1964, to the 1966 Mars launch opportunity and renamed Mariner Mars 1966.

On this sol in 204 (1993 June 08):

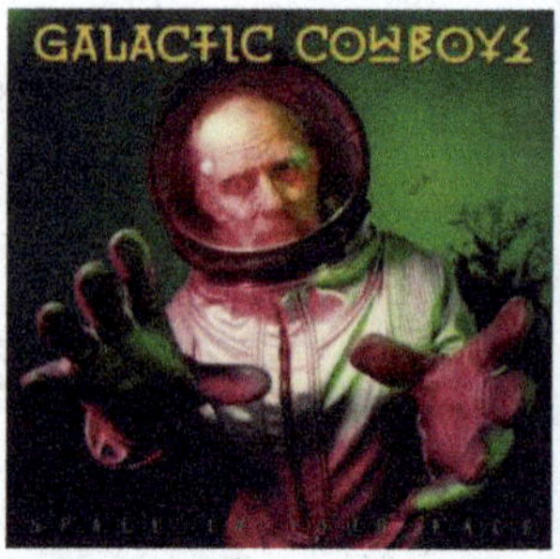

The song "Ranch on Mars" by the American heavy metal band Galactic Cowboys was released on their CD *Space in Your Face.*

> Different places, different names,
> Peculiar people just the same,
> We are brothers in our hearts,
> Though we've traveled far away,
> Another galaxy to play,
> Up in space, we never part.
>
> Someday we'll live among the stars, maybe own a ranch on Mars.
>
> Come in, come in, mission control,
> Give us guidance for our souls with eyes that scan the universe,
> No more talking, time to land,
> You give me hope, you understand,
> Our future lies beyond this Earth.
>
> Someday we'll live among the stars, maybe own a ranch on Mars.
>
> Different places, different names,
> Peculiar people just the same,
> We are brothers in our hearts.
>
> Someday we'll live among the stars, maybe own a ranch on Mars,
> Someday we'll live among the stars, maybe own a ranch on Mars.

Lyricist: Monty Colvin

On this sol in 210 (2004 September 19):

MER-B Sol 234

On this sol in 210 (2008 September 20):

MER-A Sol 255

Solar conjunction plus 5. Command sequences previously sent to *Opportunity* and *Spirit* performed a set of automatic science activities while communications with Earth were disrupted during the solar conjunction period.

—NASA/JPL

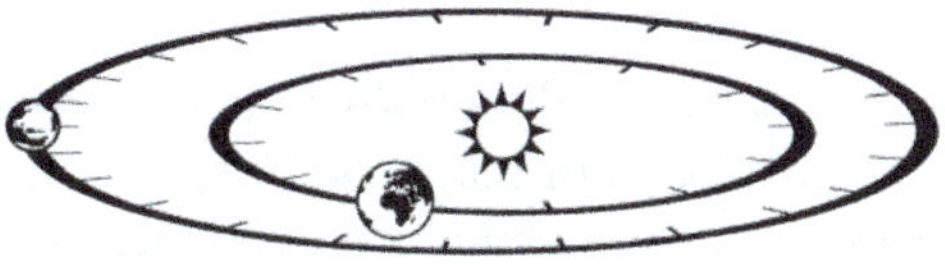

221 MINA

As seen from Mars, the Sun slowly traverses the central sector of the broad constellation Pisces as Mars descends from aphelion. The name Mina (मीना), meaning "fish" in Sanskrit, is taken from a month in the Hindu solar calendar that originally occurred when the Sun passed through the constellation Pisces as seen from Earth.

The median one-way communications delay during this month is 15 minutes 0 seconds.

221 Mina 01 **2025 May 31 05:14 UTC**

Ls 090.6° Curiosity Sol 4556
MSD 52825 Perseverance Sol 1521

Approximately the first sol of the unnamed month between December and January on the Heinlein calendar.

On this sol in 210 (2004 September 21):

MER-A Sol 256: *Spirit* took 24 more hours of Mössbauer observations on the filter magnet. A dirt clod from a previous Mössbauer soil touch was inadvertently placed on the perimeter of the filter magnet on Sol 240 (210 Pisces 13). As a result, engineers believed this could have been the rover team's last best chance to collect Mössbauer data on the uncontaminated dust sample from that magnet. This is because when the Mössbauer instrument was removed, there was a chance that dirt from the clod would sprinkle or spread to the center area of the magnet.

—NASA/JPL

221 Mina 02 **2025 June 01 05:54 UTC**

Ls 091.0° Curiosity Sol 4557
MSD 52826 Perseverance Sol 1522

On this sol in 210 (2004 September 22):

MER-A Sol 257: *Spirit* took 24 more hours of Mössbauer observations on the filter magnet.

—NASA/JPL

221 Mina 03 **2025 June 02 06:34 UTC**

Ls 091.5° Curiosity Sol 4558
MSD 52827 Perseverance Sol 1523

On this sol in 184 (1955 October 31):

Journey into Space: The World in Peril, Episode 6 airdate. The crew slow the ship down, and discuss where on Mars to land and the objectives of the trip. They regain contact with the strange voice, discovering it is a very confused Frank Rogers, ex-crew man of Freighter #2.

—Wikipedia

On this sol in 210 (2004 September 23):

MER-A Sol 258: The *Spirit* team removed the Mössbauer instrument from the filter magnet and took microscopic images of the both magnets. From the image thumbnails, the team could see that some dirt from the clod was indeed deposited on the outer area of the filter magnet. Front hazard-avoidance camera images taken after the Mössbauer spectrometer was removed clearly showed dirt still attached to the Mössbauer contact plate.

After finishing with the magnets, engineers moved the rover arm back down to the soil, to the same spot that had been touched by the Mössbauer instrument on Sol 240 (210 Pisces 13). The rover team then repeated a microscopic imager sequence of that soil to see if winds had deposited anything there during conjunction. The team then centered the alpha particle X-ray spectrometer on the same soil and started an integration later that night.

—NASA/JPL

221 Mina 04 | **2025 June 03 07:13 UTC**

Ls 091.9° | Curiosity Sol 4559

MSD 52828 | Perseverance Sol 1524

On this sol in 198 (1982 March 01):

The song "I Turned into a Martian" by the American punk rock band The Misfits was released on their vinyl LP *Walk Among Us* (date approximate). The cover art includes the Martian rat-bat-spider from the motion picture *Angry Red Planet* (1959).

See 221 Aquarius 15 for lyrics.

On this sol in 210 (2004 September 23):

MER-B Sol 238: After a well-deserved rest through solar conjunction, *Opportunity* is awake again and back to work. The conjunction was the period in mid-September when Mars was nearly behind the Sun from Earth's perspective, causing communications to be unreliable.

Opportunity completed instrument arm operations on a soil target called Auk by finishing a multi-sol Mössbauer spectrometer integration and collecting microscopic images of undisturbed soil. It then performing remote sensing observations on the next target, a rock called Ellesmere. Once the morning activities were complete, *Opportunity* took a 90-minute nap then stowed the arm and drove backwards 0.34 meters (1.1 feet). The rover used an afternoon communications session on Sol 238 and an early morning session on Sol 239.

On this sol in 210 (2004 September 24):

MER-A Sol 259: *Spirit* changed tools to the Mössbauer spectrometer and started a 24-hour integration on the same spot that had been touched by the Mössbauer instrument on Sol 240 (210 Pisces 13). *Spirit* also started a three-sol thermal investigation, using panoramic camera and miniature thermal emission spectrometer observations of soil targets several times during each sol.

—NASA/JPL

221 Mina 05 **2025 June 04 07:53 UTC**

Ls 092.4° Curiosity Sol 4560
MSD 52829 Perseverance Sol 1525

On this sol in 151 (1893 October 07):

Frederick Robert Yull (Fred Yule) was born; British actor; starred in the radio series *Journey into Space* (1954-1956).

On this sol in 210 (2004 September 24):

MER-B Sol 239: The planning session for Sols 239 and 240 was extremely challenging for the uplink team. As the rover project transitions to five-day-a-week planning, the Opportunity team planned two sols of activities to be uplinked on Sol 239. Adding to the complexity, the two sols' activities included difficult instrument arm placement activities. Rover planners rose to the occasion. The sol started with 45 minutes of microscopic imaging, then placement of the alpha particle X-ray spectrometer on a target area of Ellesmere called "No Coating." Opportunity performed a couple hours of remote sensing, used an afternoon communications session and then went into overnight deep sleep.

On this sol in 210 (2004 September 25):

MER-A Sol 260: *Spirit* completed the Mössbauer integration of the spot that had been touched by the Mössbauer instrument on Sol 240 (210 Pisces 13).

—NASA/JPL

221 Mina 06 **2025 June 05 08:32 UTC**

Ls 092.9° Curiosity Sol 4561
MSD 52830 Perseverance Sol 1526

On this sol in 158 (1906 December 09):

THERE IS LIFE ON THE PLANET MARS

By Lilian Whiting

The legions of canals on Mars, forming a colossal and a wisely planned system designed to irrigate the oases of the vast deserts which make up the surface of this planet, are an unanswerable argument for the existence of conscious, intelligent life. A thing made predicates a maker. This truism, of course, was Paley's favorite assertion, but it is none the worse for that. Schiaparelli discovered 104 canals; Prof. Percival Lowell and his staff of the Lowell Observatory at Flagstaff, Arizona, have discovered over 350. and they regard this number as no limit. The larger and more obvious are. like the larger asteroids, discovered first; but in each opposition of the planet the trained sight and skill of the great astronomer who is now held to be the specialist on Mars—the Martian expert, as it were—discovers new and smaller ones.

The observations of Mars present a series of paradoxes. Certain oppositions present the planet comparatively near, certain others more remote. But, contrary to evident expectation, the nearer approaches are by no means invariably the more favorable for observation of the canal phenomena, for the reason that in the closer approach the planet is so tilted that its southern hemisphere is the one presented, while the canals are almost entirely in the northern hemisphere.

The New York Times.

SUNDAY, DECEMBER 9, 1906.

THERE IS LIFE ON THE PLANET MARS

Prof. Percival Lowell, recognised as the greatest authority on the subject, declares there can be no doubt that living beings inhabit our neighbor world.

By Lilian Whiting.

The Peaks at Flagstaff, Arizona.

The hypothesis of canals on Mars has already emerged from its progress through the usual stages of skepticism, ridicule, and denial which every new advance in science has to encounter. It seems to be the law and the prophets regarding all phases of the conquering of the unknown. "Every generation," remarks Mrs. Julia Ward Howe, "makes a fool of the one that went before it," and, as Prof. Lowell says, "the generation that witnesses an advance is not the generation to believe in it." But a persistent incredulity may be no less stupid and dense than simple incredulity. Unreasoning denial has no more claim than unreasoning acceptance. That there are canals on Mars is to-day an accepted fact of astronomy, by any doubt of which the doubter merely argues himself unknown. The establishment of this fact is so recent as to be one of the most important truths of the new astronomy which lies well within the past half century.

—*The New York Times*, page 39

On this sol in 210 (2004 September 25):

MER-B Sol 240: *Opportunity* began taking a reading with the alpha particle X-ray spectrometer at 7:00 local solar time, then went back to sleep. After waking, it did an hour of remote sensing observations, completed the alpha particle X-ray spectrometer integration and collected more microscopic imager pictures. The rover then placed the alpha particle X-ray spectrometer on its next target, Barbeau. Another hour of remote sensing completed the sol.

On this sol in 210 (2004 September 26):

MER-A Sol 261: *Spirit* stowed its arm then drove backwards 1.5 meters (4.9 feet) to take post-conjunction panoramic camera pictures of the soil underneath the rover as part of the conjunction wind experiment. *Spirit* also took navigation camera images of the road ahead in preparation for future drives.

—NASA/JPL

221 Mina 07 **2025 June 06 09:12 UTC**

Ls 093.3° Curiosity Sol 4562

MSD 52831 Perseverance Sol 1527

On this sol in 179 (1946 June 08):

The animated short film "Kitty Kornered" was released. Porky Pig puts his cats out in the snow, but then they put him out and have a party. Expelling them again, Porky goes to bed, only to be terrorized by the felines' mock Martian invasion.

—International Movie Database

On this sol in 210 (2004 September 26):

MER-B Sol 241: *Opportunity* finished its alpha particle X-ray spectrometer integration on Barbeau, collected more microscopic images, switched tools to the Mössbauer spectrometer and started that integration. The rover performed a mini deep sleep overnight.

On this sol in 210 (2004 September 27):

MER-A Sol 262: *Spirit* successfully transitioned back to normal operations from conjunction operations, when Mars and Earth were on opposite sides of the Sun. During conjunction (Sols 244 through 255, 210 Pisces 17 through 28), engineers and scientists did not attempt normal operations due to the low probability of successful communications. From Sols 244 to 249 (210 Pisces 17 to 22), the rover team did transmit several "no operation" commands to test the communications link. On *Spirit's* Sol 249 (210 Pisces 22), the rover experienced an unexpected software reset, apparently triggered by a corrupted "no operation" command. As a result of that problem, engineers ceased all commanding on *Spirit* from Sol 250 until Sol 256 (210 Pisces 23 through 210 Mina 01), at which time the likelihood of receiving corrupted commands was once again very low.

As a result of the difficulties getting data off of the rover, the memory available for science data storage shrunk to roughly 100 megabits by Sol 261, but has recovered as of Sol 262 to roughly 400 megabits.

Due to the limited amount of available science data storage, planned activities were limited to a Mössbauer spectrometer integration on a rock, limited remote sensing, and routine atmospheric observations. That plan did not make it on board due to a problem during the communications uplink session. The deep space network antenna was pointed a few degrees below its lower safety limit when the transmitter was supposed to turn on, causing an interlock mechanism to turn off the transmitter. By the time the antenna was reconfigured, not enough time remained to get the full sequence load transmitted. Fortunately, one sequence did make it to the rover and was successfully executed, freeing up roughly 250 megabits of memory for future sols.

—NASA/JPL

221 Mina 08 **2025 June 07 09:52 UTC**

Ls 093.8° Curiosity Sol 4563
MSD 52832 Perseverance Sol 1528

On this sol in 175 (1938 December 01):

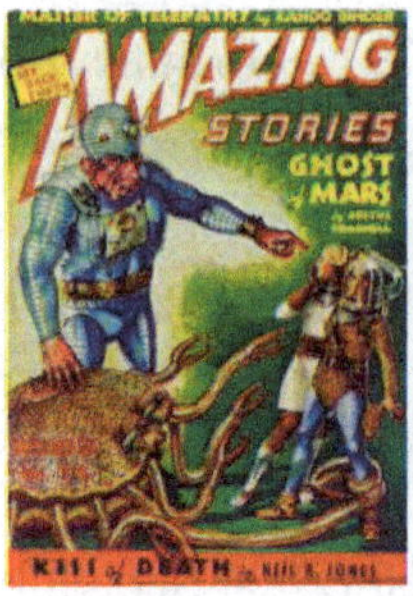

"Ghost of Mars" by Festus Pragnell appeared in Amazing Stories (date approximate). Trapped by mutinous miners, Don Hargreaves and Elsa Thorwaldson found themselves facing a grim Martian ghost deep within the tunnels of Mars.

Cover and interior art by Robert Fuqua.

On this sol in 210 (2004 September 28):

MER-A Sol 263: *Spirit* successfully drove approximately 7 meters (23 feet) and acquired images to build a digital elevation map of the hills. This put *Spirit* on the south side of a 2-meter-diameter (7-foot-diameter) depression, with Tetl on the opposite side.

—NASA/JPL

221 Mina 09 **2025 June 08 10:31 UTC**

Ls 094.2° Curiosity Sol 4564
MSD 52833 Perseverance Sol 1529

On this sol in 210 (2004 September 29):

MER-A Sol 264: *Spirit* drove about 4 meters (13 feet) around the edge of the depression to keep the rover's solar panels (which are the rover's main power source) tilted toward the Sun. Since the Sun moves low across the northern sky over Gusev Crater at this time of year, rover planners are attempting to keep the solar panels tilted toward the north. The drive included use of the five-wheel mode to minimize use of the sticky right front wheel, which inefficiently pulls too much power when it is activated. *Spirit* also gathered additional about potential science targets, using the miniature thermal emission spectrometer and panoramic camera at the end of the rover's robotic arm.

—NASA/JPL

221 Mina 10 **2025 June 09 11:11 UTC**

Ls 094.7° Curiosity Sol 4565
MSD 52834 Perseverance Sol 1530

On this sol in 184 (1955 November 07):

Journey into Space: The World in Peril, Episode 7 airdate. Thinking Frank may be there, the crew decide to land near the wreck of freighter #2, but Jet and Mitch discover it to be empty. From *Discovery,* Doc and Lemmy spot a fleet of Martian spheres approaching.

—Wikipedia

On this sol in 210 (2004 September 30):

MER-A Sol 265: *Spirit* attempted to approach Tetl, but the drive ended early because the flight software detected that a steering brake control function did not work. Remote sensing data was still acquired.

—NASA/JPL

221 Mina 11 **2025 June 10 11:50 UTC**

Ls 095.2° Curiosity Sol 4566

MSD 52835 Perseverance Sol 1531

On this sol in 151 (1893 October 13):

Hamilton Moore Jeffers was born; American astronomer; observer of Mars.

On this sol in 173 (1935 March 01):

Part 5 of *Swords of Mars* by Edgar Rice Burroughs appeared in *The Blue Book Magazine* (date approximate).

Interior art by Robert Fink.

Part 2 of The Titan by P. Schuyler Miller appeared in *Marvel Tales of Science and Fantasy* (date approximate).

On this sol in 210 (2004 October 01):

MER-B Sol 245: On this a restricted sol, Opportunity could perform only remote sensing. The rover took images in all directions with its navigation camera. It used its miniature thermal emission spectrometer for sky and ground observations. Then it went into deep sleep for the night of Sol 245 into the morning of Sol 246.

MER-A Sol 266: Engineers on NASA's Mars Exploration Rover team are investigating possible causes and remedies for a problem affecting the steering on *Spirit*.

The relay for steering actuators on *Spirit's* right-front and left-rear wheels did not operate as commanded. Each of the front and rear wheels on the rover has a steering actuator, or motor, that adjusts the direction in which the wheels are headed independently from the motor that makes the wheels roll. When the actuators are not in use, electric relays are closed and the motor acts as a brake to prevent unintended changes in direction.

Meanwhile, the team continues to use *Spirit's* robotic arm and camera mast to study rocks and soils around the rover, without moving the vehicle until the cause of the anomaly is understood and corrective measures can be implemented.

With an ongoing investigation of the steering anomaly, no further driving was planned. Several targets in front of the rover were selected for the alpha particle X-ray spectrometer and Mössbauer spectrometer. *Spirit* completed alpha particle X-ray spectrometer readings on two different locations.

—NASA/JPL

221 Mina 12 **2025 June 11 12:30 UTC**

Ls 095.6° Curiosity Sol 4567

MSD 52836 Perseverance Sol 1532

On this sol in 210 (2004 October 02):

MER-B Sol 246: *Opportunity's* next three sols were planned in a single planning cycle as part of a 5-day-a-week schedule. The uplink team accomplished a Herculean task, successfully completing and uplinking three science-intensive sol plans despite some issues encountered during the day.

Opportunity began sol 246 by placing the Mössbauer spectrometer and starting a long reading with it on a target called Void. While collecting the Mössbauer data, *Opportunity* also performed two hours of observations with its panoramic camera and its miniature thermal emission spectrometer. The Mössbauer integration was paused just before the afternoon communication session with *Mars Odyssey*. Deep sleep was disabled so that *Opportunity* could support an early morning communications session on Sol 247 and restart the Mössbauer integration.

MER-A Sol 267: *Spirit* successfully acquired remote sensing data and moved its robotic arm to put the alpha particle X-ray spectrometer at a third position during the day and a fourth position overnight.

—NASA/JPL

221 Mina 13 **2025 June 12 13:10 UTC**

L_S 096.1° Curiosity Sol 4568
MSD 52837 Perseverance Sol 1533

On this sol in 210 (2004 October 03):

MER-B Sol 247: This was Opportunity's second sol of the long Mössbauer integration; the integration ran throughout the sol until early evening, at which time *Opportunity* again paused and entered deep sleep overnight. During the day, *Opportunity* also completed a series photometric observations with its panoramic camera.

MER-A Sol 268: *Spirit* acquired additional remote sensing data and performed a tool change to the Mössb]]auer spectrometer, then started an overnight integration with that instrument.

—NASA/JPL

On this sol in 210 (2004 October 04):

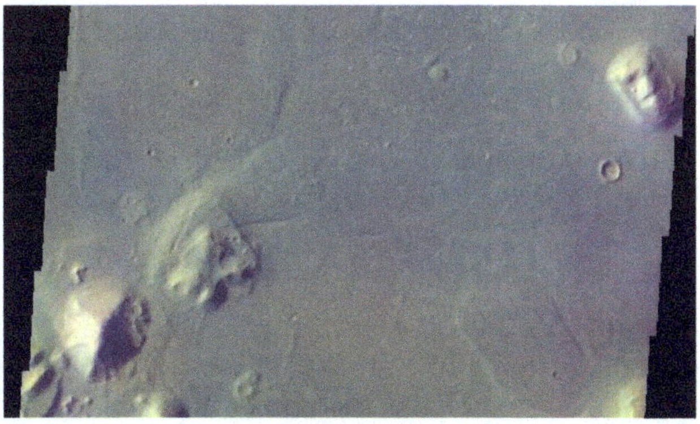

Mars Global Surveyor imaged the Cydonia Face and the DiPietro-Molenaar Pyramid.

221 Mina 14 **2025 June 13 13:49 UTC**

L_S 096.5° Curiosity Sol 4569
MSD 52838 Perseverance Sol 1534

Mars arrival opportunities from Earth via a Type I trajectory and an EEM trajectory.

On this sol in 176 (1940 October 24):

Edgar Rice Burroughs began writing "Escape on Mars," the third part of a new Mars series. The story later was published under the title "The Yellow Men of Mars."

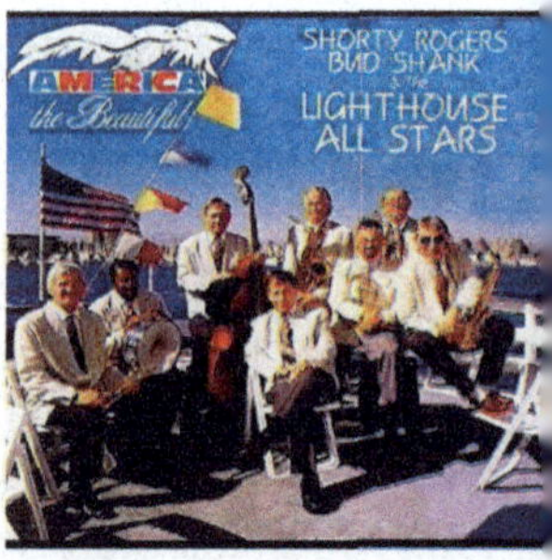

On this sol in 203 (1991 August 05):

The song "Here's That Old Martian Again" by the American jazz musicians Shorty Rogers and Bud Shank was released on their CD *America the Beautiful.*

On this sol in 210 (2004 October 04):

MER-B Sol 248: *Opportunity* exited deep sleep and restarted the Mössbauer integration. During pre-uplink science activities in the early morning, the rover completed a sky observation pattern that planners call an itty-bitty cloud movie. In the Martian afternoon, *Opportunity* ended the long Mössbauer integration and turned the tool turret on its arm to place the alpha particle X-ray spectrometer on Void. The X-ray instrument did not start taking data until early the next morning. *Opportunity* did not go into deep sleep overnight. Instead, it used an early morning *Mars Odyssey* communications session and immediately afterwards started the alpha particle X-ray spectrometer integration.

MER-A Sol 269: *Spirit* continued the Mössbauer spectrometer integration and performed a diagnostic test on the steering brake. The test indicated that there was no problem with the commanding process at that time.

—NASA/JPL

221 Mina 15 **2025 June 14 14:29 UTC**
Ls 097.0° Curiosity Sol 4570
MSD 52839 Perseverance Sol 1535

The first sol of the second quarter of Summer on the Aitken calendar.

On this sol in 155 (1901 April 26):

Fred C. Brannon was born; American film director; co-directed the serial *The Purple Monster Strikes* (1945) and the motion picture *Flying Disc Man from Mars* (1950).

On this sol in 210 (2004 October 05):

MER-B Sol 249: *Opportunity* completed the alpha particle X-ray spectrometer integration on the morning of Sol 249. The rover then performed a series of microscopic imaging activities, stowed the instrument deployment device and began driving toward its next target, Wopmay. The 19.98-meter (about 66-foot) drive went well, but with more slippage than expected. At the end of the drive, the nearest visible face of Wopmay was only 2 meters (about 7 feet) from the center of the rover; far enough not to have been a hazard during the drive, but closer than was predicted. Slip estimates indicate radial slippage as high as 64 percent. Opportunity used deep sleep overnight on Sol 249.

MER-A Sol 270: *Spirit* acquired remote sensing data, stowed its arm, and ran another steering diagnostic test. The cause of the steering brake issue has not been identified, but tests indicate that electronics related to the brake function and the overall steering capabilities of *Spirit* are healthy. Engineers are proceeding with normal operations, including mobility.

Engineers received results from *Spirit* today from a first set of diagnostic tests on the relay for steering actuators on *Spirit's* right-front and left-rear wheels. "We are interpreting the data and planning additional tests," said Rick Welch, rover mission manager at NASA's Jet Propulsion Laboratory, Pasadena, California. "We hope to determine the best work-around if the problem does persist."

JPL's Jim Erickson, rover project manager, said, "If we do not identify other remedies, the brakes could be released by a command to blow the fuse controlling the relay, though that would make those two brakes unavailable for the rest of the mission." Without the steering-actuator brakes, small bumps or dips that a wheel hits during a drive might twist the wheel away from the intended drive direction.

"If we do need to disable the brakes, errors in drive direction could increase. However, the errors might be minimized by continuing to use the brakes on the left-front and right-rear wheels, by driving in smaller segments, and by adding a software patch to reset the direction periodically during a drive," Erickson said. Engineers believe the steering-brake issue is not related to excessive friction detected during the summer in the drive motor for *Spirit's* right-front wheel, because the steering actuator is a different motor.

—NASA/JPL

221 Mina 16 **2025 June 15 15:08 UTC**

Ls 097.5° Curiosity Sol 4571
MSD 52840 Perseverance Sol 1536

On this sol in 155 (1901 April 27):

Frank Belknap Long was born; American science fiction writer; author of *Martian Homecoming* (1951), "And Someday to Mars" (1952), "The Calm Man" (1954), "The Man the Martians Made" (1954), "Good to Be a Martian" (1955), and *The Martian Visitors* (1964).

On this sol in 171 (1931 June 01):

"In the Spacesphere" by Charles Cloukey appeared in *Wonder Stories* (date approximate). The Earthman hijacks the spaceship, using his useful slug-gun.

Cover and interior art by Frank R. Paul.

On this sol in 196 (1978 June 09):

Jeff Wayne's Musical Version of The War of the Worlds was released as a set of two vinyl LP discs. Track list:

Side A
1. "The Eve of the War"
2. "Horsell Common and the Heat Ray"

Side B
1. "The Artilleryman and the Fighting Machine"
2. "Forever Autumn"
3. "Thunder Child"

Side C
1. "The Red Weed (Part 1)"
2. "Parson Nathaniel"
3. "The Spirit of Man"
4. "The Red Weed (Part 2)"

Side D
1. "Brave New World"
2. "Dead London"
3. "Epilogue (Part 1)"
4. "Epilogue (Part 2)"

On this sol in 210 (2004 October 06):

MER-B Sol 250: This image was taken by *Opportunity's* navcam inside Endurance Crater.

Opportunity performed the first part of a planned two-sol approach to Wopmay. This nearly 7-meter (23-foot) drive went well. The end of the drive incorporated conditional arcs to be executed only if the rover was in the appropriate position. The drive put the rover in very good position for the final approach on Sol 251.

MER-A Sol 271: *Spirit* successfully drove approximately 2 meters (6.6 feet). This put Tetl within reach of the robotic arm. After the drive, *Spirit* used its navigation camera to view the scene from the rover's new location.

—NASA/JPL

221 Mina 17 **2025 June 16 15:48 UTC**

Ls 097.9° Curiosity Sol 4572
MSD 52841 Perseverance Sol 1537

On this sol in 151 (1893 October 19):

Harold Vincent Schoepflin (Harl Vincent) was born; American mechanical engineer and science fiction writer; author of *The Ambassador from Mars* (1928), "Before the Asteroids" (1930), "Silver Dome" (1930), "The Terror of Air-Level Six" (1930), "Invisible Ships" (1931), "Too Many Boards" (1931), "Roadways of Mars" (1932), "Thia of the Drylands" (1932), "Lost City of Mars" (1934).

On this sol in 184 (1955 November 14):

Journey into Space: The World in Peril, Episode 8 airdate. Jet and Mitch fail to make it back to *Discovery,* and the crew all pass out as the ships approach. They wake up in a dark room and figure that they must be on one of the moons of Mars. They discover an unconscious man in the room, that Jet is missing, and a door in the floor opens.

—Wikipedia

On this sol in 210 (2004 October 07):

MER-B Sol 251: *Opportunity* has driven 1,619 meters (just over a mile). Its latest stop is a lumpy boulder dubbed Wopmay inside Endurance Crater.

The daily power supply for each rover comes from 1.3 square meters (14 square feet) of solar panels converting sunlight into electricity. Just after the landings in January, the output was about 900 watt-hours per day for each. As anticipated, output gradually declined due to dust buildup and the Martian seasonal change with fewer hours of sunlight and a lower angle of the Sun in the sky. By 210 Aquarius, *Spirit's* daily output had declined to about 400 watt-hours per day. It has been between 400 and 500 watt-hours per day for most of the past two months.

Opportunity, closer to Mars's equator and with the advantage of a sunward-facing tilt as it explored inside the southern half of a crater, maintained an output level between 500 and 600 watt- hours per day in 210 Makara, Aquarius, and Kumbha. Since early 210 Pisces, the amount of electricity from *Opportunity's* solar panels has increased markedly and unexpectedly, to more than 700 watt-hours per sol, a level not seen since the first 10 weeks of the mission (late 209 Scorpius to 210 Sagittarius).

"We've been surprised but pleased to see this increase," said Erickson, "The team is evaluating ways to determine which of a few different theories is the best explanation."

Possible explanations under consideration include the action of wind removing some dust from the solar panels or the action of frost causing dust to clump. "We seem to have had several substantial cleanings of the solar panels," Erickson said."

MER-A Sol 272: Rover engineers refrained from driving *Spirit* for five days after a Sol 266 (210 Mina 11) malfunction of a system that prevents wheels from being jostled in unwanted directions while driving. Each of the front and rear wheels of the rover has a motor called a steering actuator. It sets the direction in which the wheel is headed. The steering actuators are different from the motors that make the wheels roll, and hold the wheel in a specific direction while driving. A relay used in turning these steering actuators on and off is the likely cause of the intermittent nature of the anomaly.

The relay operates *Spirit's* right-front and left-rear wheels concurrently, and did not operate as commanded on 210 Mina. Subsequent testing showed no trace of the problem, and on 210 Mina 17 the rover steered successfully and drove about 2 meters (7 feet), putting it in position to examine a layered rock called Tetl for several days.

Spirit took images with the microscopic imager to create a mosaic of Tetl's layered rock face.

—NASA/JPL

221 Mina 18 **2025 June 17 16:27 UTC**

Ls 098.4° Curiosity Sol 4573
MSD 52842 Perseverance Sol 1538

On this sol in 210 (2004 October 08):

MER-B Sol 252: *Opportunity's* Sols 252 through 254 were planned as a group, due to the rover team's switch to a five-day-a-week planning schedule. On the morning of Sol 252, *Opportunity* performed remote sensing, including two photometry survey observations, a cloud-search movie and imaging of Wopmay with the panoramic camera using 13 filters. In the afternoon, the rover backed away from the target to reach a staging position for attempting to reach the most interesting part of Wopmay, the upper right lobe, on a drive after the rover team's weekend. Deep sleep was disabled, allowing *Opportunity* to support a *Mars Odyssey* pass in the early morning of Sol 253.

MER-A Sol 273: *Spirit* captured more microscopic images of Tetl's layered face, then put the alpha particle X-ray spectrometer in place for an early morning observation.

—NASA/JPL

221 Mina 19 **2025 June 18 17:07 UTC**

Ls 098.9° Curiosity Sol 4574
MSD 52843 Perseverance Sol 1539

On this sol in 210 (2004 October 09):

MER-B Sol 253: *Opportunity's* Sols 253 and 254 were remote-sensing sols. On Sol 253, the rover imaged its filter magnets and performed multiple observations with its miniature thermal emission spectrometer. Before the rover shut down for the night, it positioned the spectrometer for a middle-of-the-night observation.

MER-A Sol 274: *Spirit* woke up at 4:00 a.m. to start the alpha particle X-ray spectrometer observation. The alpha particle X-ray spectrometer stayed on until the start of normal morning atmospheric science observations. *Spirit* also used its miniature thermal emission spectrometer to observe nearby rocks named Zackuk and Palenque, which are possible future targets for in-depth observations. Later, *Spirit* changed tools on its robotic arm, placing the Mössbauer instrument on Tetl for an observation the next morning.

—NASA/JPL

221 Mina 20 **2025 June 19 17:47 UTC**

Ls 099.3° Curiosity Sol 4575

MSD 52844 Perseverance Sol 1540

On this sol in 204 (1993 June 29):

Jazz musicians Shorty Rogers and Bill Perkins recorded the songs "Martians Go Home" and "Have You Hugged Your Martian Today" at the Royal Palms Inn in Phoenix, Arizona.

On this sol in 210 (2004 October 10):

MER-B Sol 254: *Opportunity* woke up first at 10 p.m. local solar time to begin heating, then again at 11 p.m. for the actual observation, then slept until about 7 a.m. local solar time. Upon waking, the rover began another day of remote science, including multiple observations of sky and ground with its miniature thermal emission spectrometer.

MER-A Sol 275: *Spirit* completed a 6-hour Mössbauer integration and performed daily atmospheric observations. This was the final sol of *Spirit's* weekend plan and was purposely simple to enable the sequencing team to complete a 3-day plan on Friday.

—NASA/JPL

221 Mina 21 **2025 June 20 18:26 UTC**

Ls 099.8° Curiosity Sol 4576

MSD 52845 Perseverance Sol 1541

On this sol in 141 (1875 January 01):

Mabel Winifred Knowles (Lester Lurgan) was born; British writer; author of *A Message from Mars: A Story Founded on the Popular Play by Richard Ganthony* (1912).

On this sol in 147 (1886 April 14):

Robert F. Hill was born; American film director; co-directed the serial motion picture *Mars Attacks the World* (1938).

On this sol in 210 (2004 October 11):

MER-B Sol 255: During planning for Sol 255 it was decided that approaching Wopmay from directly uphill would be too risky for *Opportunity* due to concerns about vehicle stability, so the plan was revised to approach Wopmay very slowly from an angle and not to deploy the robotic arm with any pre-loading on the rock. This would minimize chances of the vehicle slipping or placing undue stress on the arm. So the plan for Sol 255 was to back up the hill, away from Wopmay, and position *Opportunity* to begin the slow angular, downhill approach to Wopmay. The drive successfully covered 7.8 meters (about 26 feet), leaving Wopmay 3.4 meters (11 feet) away. *Opportunity* also performed about an hour of remote science observations before entering deep sleep for the night.

MER-A Sol 276: *Spirit* restarted the Mössbauer instrument at 4:00 a.m. for another 10 hours of integration time on the same spot. *Spirit* also took a few final microscopic images of Tetl, then stowed the robotic arm in preparation for the next sol's drive.

—NASA/JPL

221 Mina 22 **2025 June 21 19:06 UTC**
Ls 100.2° Curiosity Sol 4577
MSD 52846 Perseverance Sol 1542

On this sol in 196 (1978 June 15):

The song "Twelve Bars from Mars" by the American jazz fusion musician Lenny White was released as a track on his vinyl LP *Streamline.*

On this sol in 210 (2004 October 12):

MER-B Sol 256: *Opportunity* continued its slow approach toward Wopmay. After receiving its morning uplink, *Opportunity* performed twenty minutes of panoramic camera observations, then crawled forward just over 2 meters (about 7 feet), taking three small steps and performing a turn in place. During these moves *Opportunity* experienced slippage as high as 54 percent. By the end of the drive, Wopmay was approximately 1.5 meters (about 5 feet) away. *Opportunity* performed another hour of post-drive remote science observations, took a nap, then used an afternoon communication session and early morning communications with *Mars Odyssey.*

MER-A Sol 277: *Spirit* attempted a drive to the next rock target, Uchben (which means "ancient" in the old Mayan language), another layered rock roughly 6 meters (20 feet) northeast of Tetl. Halfway into that drive, *Spirit* experienced a repeat problem in the steering motor control system that engineers first saw on Sol 265 (210 Mina 10). The mobility software attempted to move a steering motor by first commanding open a relay (electronic switch) that releases a dynamic brake. The feedback from that command indicated that the relay was still closed, so the motor control software declared an error. Due to the error, the rover ignored that steering command and all subsequent driving commands. The root cause of the failed relay command is under investigation.

—NASA/JPL

221 Mina 23 **2025 June 22 19:45 UTC**
Ls 100.7° Curiosity Sol 4578
MSD 52847 Perseverance Sol 1543

On this sol in 210 (2004 October 13):

MER-B Sol 257: During the morning *Opportunity* extended its arm to use its microscopic imager for an hour on soil targets just in front of Wopmay. The rover then stowed its arm and began the final approach toward Wopmay. The drive was commanded for rolling 30 centimeters (about 1 foot) with the expectation of approximately 50-percent slippage. The expected traverse was 45 centimeters (about 1.5 feet). The last few drives taxed the rover planners and they really came through. *Opportunity* is perfectly positioned to examine Wopmay with the rover's microscopic imager and alpha particle X-ray spectrometer.

MER-A Sol 278: Engineers repeated diagnostic tests for *Spirit's* steering problem. Those tests showed that the electronics relay in question is still functional, but appears to operate intermittently. *Spirit* is otherwise healthy and is in a safe location.

A diagnostic test last run on Sol 270 (210 Mina 13) was repeated. That test showed that the steering motor's dynamic brake relay can still be opened and closed, but does occasionally (5 out of 10 times) indicate that it is still closed after being commanded open.

Future plans are to clear the drive error and attempt another drive to Uchben on Sol 281 (210 Mina 26). Engineers are also planning to run more diagnostic tests starting on Sol 282.

More diagnostics tests are needed before the source of the problem can be positively identified. Until then, engineers will continue to drive, but will steer the rover in a tank-like fashion, not using the steering actuator in question.

—NASA/JPL

221 Mina 24 **2025 June 23 20:25 UTC**

Ls 101.2° Curiosity Sol 4579
MSD 52848 Perseverance Sol 1544

On this sol in 184 (1955 November 21):

Journey into Space: The World in Peril, Episode 9 airdate. Jet is found under his bed and the unconscious man is found to be Frank Rogers. Doc tries to bring Frank out of his conditioned state, and he reveals that the invasion is under way. Jet and Lemmy go through the floor, and are directed by the voice of Paddy Flynn.

—Wikipedia

On this sol in 210 (2004 October 14):

MER-B Sol 258: *Opportunity* examined three targets (Otter, Jenny, and Hiller) on Wopmay with its microscopic imager. Each of the observations was designed to produce a mosaic. They relied on touching the instrument's contact sensor to the uneven surface of Wopmay for each quadrant of each mosaic to ensure appropriate standoff distances for good focus.

MER-A Sol 279: *Spirit* was parked at the location where a second occurrence of a problem with the rover's dynamic brake relay anomaly had halted a planned drive on Sol 277 (210 Mina 22). Scientists took the opportunity to analyze disturbed soil in front of the rover. *Spirit* deployed its robotic arm, acquired images of the soil with the microscopic imager, and placed the alpha particle X-ray spectrometer on a new target, named TakeABreak, for an overnight integration.

—NASA/JPL

221 Mina 25 **2025 June 24 21:05 UTC**

Ls 101.6° Curiosity Sol 4580
MSD 52849 Perseverance Sol 1545

On this sol in 155 (1901 May 07):

Albert DeMond was born; American screenwriter, producer, and director; co-wrote the screenplay for the motion picture *The Purple Monster Strikes* (1945).

On this sol in 210 (2004 October 15):

MER-B Sol 259: Sols 259 through 261 were designed as a single three-sol weekend plan for *Opportunity*. During the morning of Sol 259, arm operations continued with two more microscopic imager mosaics of the targets Jet Ranger 2 and Twin Otter. In the early afternoon, *Opportunity* placed its alpha particle X-ray spectrometer in a hover position approximately 1 centimeter (0.4 inches) above Otter. After a couple of naps and miniature thermal emission spectrometer observations, the rover went into deep sleep until the next morning.

MER-A Sol 280: Sols 280, 281, and 282 were built as a single three-sol plan for *Spirit* to execute over Earth's weekend. On Sol 280, *Spirit* acquired morning observations of sky and ground with its miniature thermal emission spectrometer, took a panoramic camera image to assess atmospheric quality, and completed the overnight alpha particle X-ray spectrometer measurement. After a midday nap, *Spirit* did a tool change from the alpha particle X-ray spectrometer to the Mössbauer spectrometer and began an overnight integration on the same soil patch.

—NASA/JPL

221 Mina 26 **2025 June 25 21:44 UTC**

Ls 102.1° Curiosity Sol 4581
MSD 52850 Perseverance Sol 1546

On this sol in 139 (1871 April 02):

Joseph W. Girard was born; American actor; starred in the motion picture *Nan of the North: Missil from Mars* (1922).

On this sol in 210 (2004 October 16):

MER-B Sol 260: The first part of a reading with the alpha particle X-ray spectrometer over Otter was performed for three hours in the morning. During the same period, the rover made navigation camera, panoramic camera, and miniature thermal emission spectrometer cloud observations, then snapped images with the navigation and panoramic cameras in the drive direction. While performing an atmospheric observation with the panoramic camera, *Opportunity* collected extra images of the sky close to the Sun to allow observations of the dust accumulation on the camera's window.

MER-A Sol 281: *Spirit* completed the Mössbauer spectrometer measurement, took a midday nap, acquired three images of a nearby target called Coffee with the microscopic imager, and stowed the robotic arm. *Spirit* then successfully drove about 4 meters (13 feet) backwards, putting the target Uchben into the workspace of the robotic arm. The drive included straightening the right front and left rear steering wheels, which are the two impacted by a problem with the relay that is used in turning the steering motors on and off. The drive also successfully tested driving without use of the right front and left rear steering wheels to limit use of these motors while investigation of the malfunction continues.

—NASA/JPL

221 Mina 27 **2025 June 26 22:24 UTC**
Ls 102.6° Curiosity Sol 4582
MSD 52851 Perseverance Sol 1547

On this sol in 210 (2004 October 17):

MER-B Sol 261: In the early morning, the reading with *Opportunity's* alpha particle X-ray spectrometer resumed, continuing until about 11 a.m. local solar time and producing excellent spectra despite the standoff position of the sensor. *Opportunity* then collected Mössbauer data over Otter for another three hours. Several remote science observations were made over the course of the sol, including photometric measurements with the panoramic camera and targeted observations of Wopmay with the miniature thermal emission spectrometer and the panoramic camera observations.

MER-A Sol 282: *Spirit* acquired measurements of the sky and ground in the morning with the miniature thermal emission spectrometer, took the usual midday nap, and then made remote-sensing observations in the afternoon, including some navigation camera images for use in planning of future driving.

—NASA/JPL

221 Mina 28 **2025 June 27 23:03 UTC**
Ls 103.0° Curiosity Sol 4583
MSD 52852 Perseverance Sol 1548

On this sol in 172 (1933 May 01):

"Martian and Troglodyte" by Neil R. Jones appeared in *Amazing Stories* (date approximate). The two Martians were all eyes, for the wonders of their neighboring world riveted their attention, attracting them like a magnet. Nearer and nearer to the surface they dropped.

Interior art by Leo Morey.

A Fighting Man of Mars by Edgar Rice Burroughs was republished (date approximate). Hadron of Hastor, Padwar in the service of the Warlord of Mars, goes alone in search of the beautiful Sanoma Tora when she is abducted by Tul Axtar of Jahar. But before Hadron manages to reach Jahar he is deflected from his purpose in an effort to rescue a slave girl of his own race from the ferocious green men of mars. Tavia, for that is the lovely girl's name, leads Hadron to her home city Tjanath from which she had been stolen as a very small girl.

Cover art by J.H. Hartley.

On this sol in 210 (2004 October 18):

MER-B Sol 262: In the morning, *Opportunity* took a third photometric measurement and used the panoramic camera to study Wopmay, concluding the weekend plan. In the early afternoon, the rover made a final microscopic image of the target Hercules on Wopmay, then backed away as part of a plan to re-approach the back side of Wopmay for possible additional measurements. Navigation camera imagery of the back of Wopmay was collected in the afternoon, followed by extensive panoramic camera imaging the next morning. Unfortunately, due to slippage during the traverse, *Opportunity* didn't reach its desired vantage point, and the target was not in view.

MER-A Sol 283: After receiving its daily commands and acquiring a panoramic camera assessment of atmospheric quality and miniature thermal emission spectrometer measurements of the sky and ground, *Spirit* took a midday nap. In the afternoon, *Spirit* deployed the robotic arm and acquired 20 images of a target region called Koolik on Uchben with the microscopic imager. *Spirit* then deployed the alpha particle X-ray spectrometer for an overnight integration.

—NASA/JPL

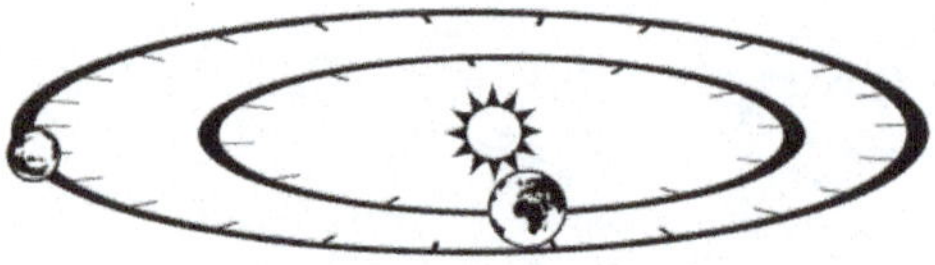

221 ARIES

As seen from Mars, the Sun slowly traverses the eastern sector of the broad constellation Pisces, entering the western sector of the constellation Aries later in the month. Originally depicted by the Babylonians as ^{MUL}LÚ.ḪUN.GÁ, the agrarian laborer, in a tradition dating from the Middle Bronze Age (4,100 to 3,550 years before present), the latter constellation's association shifted over the centuries to that of a shepherd and eventually to that of a ram.

The median one-way communications delay during this month is 16 minutes 42 seconds.

221 Aries 01 **2025 June 28 23:43 UTC**

Ls 103.5° Curiosity Sol 4584
MSD 53853 Perseverance Sol 1549

Approximately the first sol of January on the Heinlein calendar.

Earth departure opportunity to Mars via an EEM trajectory.

Mars arrival opportunity from Earth via an EEM trajectory.

Earth arrival opportunity from Mars via a Type IV trajectory.

On this sol in 118 (1831 October 07):

Eleanor Maria Easterbrook Ames (Eleanor Kirk) was born; American writer, author of *The Christ of the Red Planet* (1901). The book was written in the first person as a tale of astral projection to Mars, although she did not unambiguously identify herself as a spiritualist, and she was neither a séance medium nor a trance medium, in contrast to Sara Weiss and Catherine-Elise Müller (Hélène Smith) who both claimed to have had actual experiences on Mars.

On this sol in 171 (1931 June 15):

"The Man from Mars" by P. Schuyler Miller appeared in *Wonder Stories Quarterly*. Morally superior Martian runs afoul of the vagaries of human behavior.

Interior art by Marco Marchioni.

"The Great Invasion" by Sidney Patzer also appeared in *Wonder Stories Quarterly*. Atmospheric disturbances, and a great green bowl-like energy phenomenon closes over a part of Ohio. Plane reconnaissance shows that within the bowl are three gigantic ships. Hostilities begin almost immediately.

—Bleiler and Bleiler, 1998

Interior art by Marco Marchioni.

"Outcast in Space" by Arthur G. Stangland also appeared in *Wonder Stories Quarterly.* Science and technology have advanced considerably, with good television, and many personal planes. Interplanetary travel to Mars is common, and Martian and Terrestrial capitalists vie to create huge cartels. The Martians, who are entirely human, have a considerable establishment on Earth. To writer James Hamlyn Cord's woodland house comes a beautiful young damsel in distress.

—Bleiler and Bleiler, 1998

Interior art by Marco Marchioni.

On this sol in 210 (2004 October 19):

MER-B Sol 263: A decision was made to delete *Opportunity's* panoramic camera imagery without downlinking it, since on board memory was tight and we already have complete coverage of the part of Wopmay captured. The new plan was to continue the rover's drive toward the back of Wopmay, and repeat the imaging observations attempted in the Sol 262 plan. But as *Opportunity* drove toward Wopmay, it encountered a hidden obstacle: a rock buried under the sand that resulted in 100 percent slip for a good part of the traverse. Once the rover was clear of the rock, it continued from a point much closer to Wopmay than anticipated. Rover planners estimated that, at the end of the traverse, *Opportunity* was within 30 centimeters (just under one foot) of Wopmay, uncomfortably close.

On this sol in 210 (2004 October 20):

MER-A Sol 284: *Spirit* completed the alpha particle X-ray spectrometer measurement and then did a tool change to the Mössbauer spectrometer for a nighttime integration on Koolik.

—NASA/JPL

221 Aries 02 **2025 June 30 00:23 UTC**

Ls 104.0° Curiosity Sol 4585

MSD 53854 Perseverance Sol 1550

On this sol in 184 (1955 November 28):

Journey into Space: The World in Peril, Episode 10 airdate. Frank regains some of his memory, revealing that the crew are aboard an asteroid ship. Jet and Lemmy meet Paddy face to face, and he reveals that he is the rebel captain of the asteroid, working with Jack Evans. Paddy receives orders to have the asteroid join the invasion fleet.

—Wikipedia

On this sol in 210 (2004 October 20):

MER-B Sol 264: *Opportunity* backed away from Wopmay. The planned drive was 2.5 meters (8.2 feet) but the actual drive was 2.57 meters (8.4 feet). *Opportunity* is now in position to begin its drive toward Burns Cliff.

Total odometry after Sol 264 is 1,638.57 meters (1.0181 miles).

On this sol in 210 (2004 October 21):

MER-A Sol 285: *Spirit* continued systematic atmospheric observations on this and all sols during this period using the miniature thermal emission spectrometer and panoramic camera. The rock abrasion tool was employed to drill a shallow hole at Koolik, a location on the rock Uchben.

—NASA/JPL

221 Aries 03 **2025 July 01 01:02 UTC**
Ls 104.4° Curiosity Sol 4586
MSD 53855 Perseverance Sol 1551

On this sol in 153 (1897 August 07):

Sewell Peaslee Wright was born; American writer, author of *The Martian Missile* (1960).

On this sol in 210 (2004 October 21):

MER-B Sol 265: *Opportunity* began its drive away from a boulder called Wopmay. The rover performed 45 minutes of remote observations and then attempted a 21-meter (69-foot) drive away from Wopmay. The drive stopped after 3.5 meters (11.5 feet). *Opportunity* experienced a drive and mobility goal error due to high current draw in the steering motors.

On this sol in 210 (2004 October 22):

MER-A Sol 286: *Spirit* took microscopic images of the Koolik rock abrasion tool hole and placed the alpha particle X-ray spectrometer on Koolik for an overnight observation. The alpha particle X-ray spectrometer works best when cold.

"We are continuing tests on *Spirit* and in our testbed here at JPL," said Jim Erickson, Mars Exploration Rover project manager at JPL. One possible work-around would be to deliberately blow a fuse controlling the relay, disabling the brake action of the steering actuators. The rovers could be operated without that feature. "The only change might be driving in shorter steps when the rover is in rugged terrain," Erickson said.

Spirit has driven a total of 3,647 meters (2.27 miles) since landing, more than six times the distance set as a goal for the mission. Its current target is a layered rock called Uchben in the Columbia Hills.

—NASA/JPL

221 Aries 04 **2025 July 02 01:42 UTC**
Ls 104.9° Curiosity Sol 4587
MSD 53856 Perseverance Sol 1552

On this sol in 210 (2004 October 22):

MER-B Sol 266: Due to *Opportunity's* heading at the end of Sol 265, the morning uplink session was occluded by the panoramic camera, raising concern that we might fail to get the 3-sol command load to the spacecraft. To avoid this problem, the team instead chose to implement a high-priority communication window at 11:30 local solar time and to uplink all sequences at that time, activating the Sol 266 master sequence by real-time command. This plan worked as designed, and all sequences got onboard.

The original plan for Sols 266 and 267 was to place the alpha particle X-ray spectrometer on the filter magnet for extended integrations. However, *Opportunity's* position against a buried rock (informally named Son of Bane) and the churning up of sandy terrain meant that engineers could not rule out the possibility of an unseen rock in the robotic arm's work volume. As a consequence, the arm deployment was cancelled, and the activities for Sols 266-267 were limited to remote-sensing observations.

On this sol in 210 (2004 October 23):

MER-A Sol 287: *Spirit* placed the Mössbauer spectrometer on the Koolik rock abrasion tool hole for several observations over the Earth weekend. The Mössbauer spectrometer radiation source has weakened significantly since landing, through normal decay, so longer integration times are now required to get acceptable data.

—NASA/JPL

221 Aries 05 **2025 July 03 02:21 UTC**
Ls 105.4° Curiosity Sol 4588
MSD 53857 Perseverance Sol 1553

On this sol in 210 (2004 October 23):

MER-B Sol 267: Because of *Opportunity's* position against a buried rock (informally named Son of Bane) and the churning up of sandy terrain meant that engineers could not rule out the possibility of an unseen rock in the robotic arm's work volume, the arm deployment was cancelled, and the activities were limited to remote-sensing observations.

On this sol in 210 (2004 October 24):

MER-A Sol 288: *Spirit* continued integration of the Mossbauer spectrometer on the Koolik rock abrasion tool hole.

—NASA/JPL

221 Aries 06 **2025 July 04 03:01 UTC**
Ls 105.8° Curiosity Sol 4589
MSD 53858 Perseverance Sol 1554

On this sol in 210 (2004 October 24):

MER-B Sol 268: *Opportunity* drove away from Son of Bane. The rover turned and drove forward a short distance to get out of the hole it had dug for itself. It drove about 4.5 meters (about 15 feet) cross slope, and then began an up-slope drive designed to cover 9 meters (29.5 feet). Only about 0.4 meter (1.3 feet) of this last leg was achieved before the rover again encountered 100 percent slip.

On this sol in 210 (2004 October 25):

MER-A Sol 289: *Spirit* continued integration of the Mössbauer spectrometer on the Koolik rock abrasion tool hole.

—NASA/JPL

221 Aries 07 **2025 July 05 03:40 UTC**

Ls 106.3° Curiosity Sol 4590

MSD 53859 Perseverance Sol 1555

On this sol in 183 (1954 January 15):

Mars Clock in Debut

Device is Down to Earth — It Can Be Used in Photo Work

The world's first inter-planetary timepiece was unveiled by the Hamilton Watch Company last night at the Waldorf-Astoria Hotel, and, oddly enough, it actually has an immediately practical scientific use. The clock simultaneously records the hour, day, month and year on Earth and Mars.

The inventor of the clock is Dr. I. M. Levitt, director of the Franklin Institute's Fels Planetarium in Philadelphia. The clock can be used, he explained, when making sequence photographs on Mars. A device for measuring Martian time makes it possible to tell at what point each photograph was taken. The principles involved can be applied to timepieces geared to other planets, he said.

Other uses of the clock lie in interplanetary communication and timing in space travel in twenty years or so. Dr. Levitt admitted that he did not foresee space travel before that time, and so, the first year that appears on the clock's face is 1970. The clock, which has a face sixteen inches across, has four systems of dials.

—*The New York Times,* page 21

On this sol in 210 (2004 October 25):

MER-B Sol 269: Due to the large slippage and unsuccessful drive on Sol 268 (210 Aries 06), this sol was used to take detailed images of *Opportunity's* position and to allow the rover mobility team to plan drive strategies for subsequent sols. *Opportunity* performed more than two hours of remote observations. The rover began a routine of forfeiting deep sleep for as many sols as the battery state of charge would allow. Solar exposure has been favorable enough to reduce the need for deep sleep. In an effort to reduce the backlog of onboard science telemetry that has not been downlinked, *Opportunity* will support early morning *Mars Odyssey* communication sessions as long as the battery state of charge is not impaired.

On this sol in 210 (2004 October 26):

MER-A Sol 290: *Spirit* performed tests to diagnose the root cause of the indication that steering brakes had failed to release, but the tests were inconclusive. *Spirit* then used the rock abrasion tool to brush Chiikbes, another location on Uchben. *Spirit* placed the alpha particle X-ray spectrometer on Chiikbes for an overnight observation.

—NASA/JPL

221 Aries 08 **2025 July 06 04:20 UTC**

Ls 106.8° Curiosity Sol 4591

MSD 53860 Perseverance Sol 1556

On this sol in 150 (1891 December 21):

Malcolm Routh Jameson was born; American naval officer and science fiction writer; author of "Blockade Runner" (1941).

On this sol in 210 (2004 October 26):

MER-B Sol 270: This was the first sol dedicated to a potpourri of mobility maneuvers to gain a better understanding of the terrain on which *Opportunity* is driving. The rover completed the drive with no errors and achieved a positive elevation change of more than a meter (3.3 feet). Driving at a 45-degree angle to the slope appears to be the most productive operation.

On this sol in 210 (2004 October 27):

MER-A Sol 291: *Spirit* employed its full instrument suite on Sols 285 through 291 to study Uchben, an interesting rock encountered on the way into the Colombia Hills. The engineering team continued to diagnose and study work-arounds for a problem with the steering brake relay. An anomaly related to electric-circuit grounding came to light during this period and is also being studied by the engineering team. Neither problem has hampered *Spirit's* daily operations. *Spirit* is otherwise healthy and ready to continue its trek further into the Columbia Hills.

The engineering team has been studying recovery options for steering brakes that apparently failed to release on two previous sols. On *Spirit* and *Opportunity,* dynamic braking is accomplished using a relay switch to place a short across the motor windings of an actuator that is not being used. If that actuator starts to move unexpectedly, the motor acts as a generator and the short provides an electrical load that slows the motor down. The same principle is used to generate electrical energy for hybrid cars when the brakes are applied. Thanks to forethought on the part of the rover design team, it is possible to disable the dynamic braking function using ground commands. Those commands deliberately and safely blow a fuse that is in line with the brake relay circuit. The absence of the braking function for the steering actuators in question (right front and left rear steering) will not affect the accuracy of our drives or the rover's safety when we are stopped. Until this problem is fully resolved, we will continue to drive with the right front and left rear steering actuators disabled, using tank-like steering.

Regarding the grounding anomaly, the engineering team regularly receives telemetry that tells them the voltage difference between "rover chassis" and "power bus return". The rover chassis is the conductive structure of the rover akin to an automobile chassis. The power bus return is a collection of wires designed to carry current back to the rover power source (battery or solar array). Ideally, all rover current flows in a loop from the battery or solar array, returning by way of the power bus return wires. No current is supposed to flow in the rover chassis though, in reality, some leakage paths exist that allow current to return by way of the rover chassis. When these currents flow across the circuitry that separates the rover chassis and power bus return, they create a small voltage that is measured and reported in telemetry. Until Sol 287, the reported voltage was typically in the range of 0.6 to 0.8 volts. On Sol 287, that voltage dropped to 0 volts. The 0 volt reading could indicate that there is a problem with the measurement circuit, or it could indicate that power bus return and rover chassis are now shorted (making direct contact). The rover can operate when the chassis and power bus return are shorted together or when they are separated from each other by electrical circuitry. In the shorted case, however, the rover is more susceptible to permanent damage if another short occurs somewhere else. Engineers are looking at when the short indication occurred for clues about its possible root cause.

—NASA/JPL

221 Aries 09 **2025 July 07 05:00 UTC**

Ls 107.2° Curiosity Sol 4592

MSD 53861 Perseverance Sol 1557

The first sol of August, marsrev 3679 on the Levitt calendar.

On this sol in 148 (1888 March 18):

Baron Sandor "Alexander" Leydenfrost was born; Hungarian-American magazine artist; created the interior art for "The Million-Year Picnic" by Ray Bradbury (1946).

On this sol in 184 (1955 December 05):

Journey into Space: The World in Peril, Episode 11 airdate. Paddy leaves the crew, who learn how to manipulate the "visionphone." As Mitch and Frank follow them along a circular corridor using the cameras, Jet, Lemmy and Doc hear a voice they assume to be a Martian, ordering them back.

—Wikipedia

On this sol in 210 (2004 October 27):

MER-B Sol 271: With enthusiasm running high, the uplink team employed strategies of the drive from Sol 270 (210 Aries 08) to plan Sol 271's drive. *Opportunity* was to drive up-slope at an angle heading east, towards Burns Cliff, as part of a longer switchback drive operation. But as has often been the case recently, the drive was not successful. *Opportunity* moved 0.78 meters in a beeline (about 2.6 feet) but experienced 100 percent slippage for most of the drive and ended up approximately 0.35 meters (1.1 feet) lower.

The result of this drive calls into question *Opportunity's* ability to reach Burns Cliff with the current approach. The team is assessing other possibilities.

Total odometry after Sol 271 is 1664.43 meters (1.03 miles).

—NASA/JPL

221 Aries 10 — **2025 July 08 05:39 UTC**

Ls 107.7° — Curiosity Sol 4593

MSD 53862 — Perseverance Sol 1568

On this sol in 176 (1940 November 18):

Edgar Rice Burroughs began writing "The Invisible Men of Mars," the fourth part of a new Mars series.

221 Aries 11 — **2025 July 09 06:19 UTC**

Ls 108.2° — Curiosity Sol 4594

MSD 53863 — Perseverance Sol 1569

On this sol in 136 (1865 August 25):

A 160-million-year-old, approximately 5-kilogram shergottite meteorite designated Shergotty was witnessed to fall near Shergahti, Bihar State, India. It was later determined to have originated on Mars.

On this sol in 174 (1937 February 14):

The animated short film "Scrappy's Trip to Mars" was released. Scrappy and Oppie dream that they take a rocket trip to Mars. There they meet some nice Martian girls. When they want to go home, the Martians will not let them leave the planet.

221 Aries 12 — **2025 July 10 06:58 UTC**

Ls 108.7° — Curiosity Sol 4595

MSD 53864 — Perseverance Sol 1570

On this sol in 158 (1907 January 13):

"That there may be types of life of some kind on Mars is, I should think, quite likely."

—Sir Robert S. Ball, *The New York Times,* page 40

"Life not wholly unlike that on the earth may exist upon Mars for anything we know to the contrary."

—Simon Newcomb, *The New York Times,* page 40

On this sol in 149 (1890 February 06):

Mentore Maggini was born; Italian astronomer, observer of Mars.

On this sol in 155 (1901 May 22):

Ed Earl Repp was born; American science fiction writer; author of "Martian Terror" (1940).

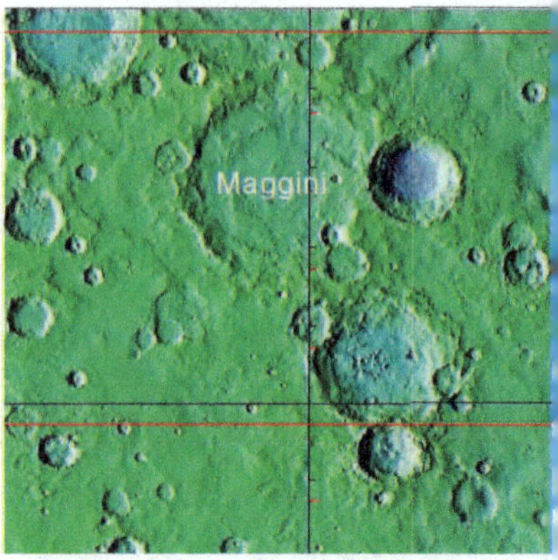

143.0-kilometer diameter crater located at 28.0°N, 9.4°E

221 Aries 13 **2025 July 11 07:38 UTC**

Ls 109.1° Curiosity Sol 4596
MSD 53865 Perseverance Sol 1571

On this sol in 173 (1935 April 01):

"The Martian Mail" by J. Lewis Burtt appeared in *Amazing Stories* (date approximate). The new mail ship, the *Marterra*, operates on a novel (apparently untested) principle. An apparatus on board changes the nature of space before it, creating a gravitational pull that drags it along. The trip proceeds smoothly for a time, but then the chief engineer worriedly informs the captain and the president of the line that the gravitational vortex has gotten out of control and that the ship is in danger.

—Bleiler and Bleiler, 1998

Interior art by Leo Morey.

Part 6 (the conclusion) of *Swords of Mars* by Edgar Rice Burroughs appeared in *The Blue Book Magazine* (date approximate).

221 Aries 14 **2025 July 12 08:18 UTC**

Ls 109.6° Curiosity Sol 4597
MSD 53866 Perseverance Sol 1572

On this sol in 150 (1891 December 27):

John Grant was born; American screenwriter, producer, and actor; co-wrote the screenplay for the motion picture *Abbott and Costello Go to Mars* (1953).

On this sol in 152 (1895 October 02):

William Alexander "Bud" Abbott was born; American actor and producer; starred in the motion picture *Abbott and Costello Go to Mars* (1953).

On this sol in 180 (1948 June 01):

"…And the Moon Be Still as Bright" by Ray Bradbury appeared in *Thrilling Wonder Stories* (date approximate). An archaeologist on Mars goes berserk and takes the side of the extinct Martian race.

—International Movie Database

Interior art by Virgil Finlay.

221 Aries 15 **2025 July 13 08:57 UTC**
Ls 110.1° Curiosity Sol 4598
MSD 53867 Perseverance Sol 1573

On this sol in 210 (2004 November 03):

MER-A Sol 298: Between Sols 292 and 298, *Spirit* completed its studies of the rock called Uchben and drove west about 2 meters (almost 7 feet) to a rock called Lutefisk.

—NASA/JPL

221 Aries 16 **2025 July 14 09:37 UTC**
Ls 110.5° Curiosity Sol 4599
MSD 53868 Perseverance Sol 1574

On this sol in 167 (1923 December 21):

Karl Stephan was born; German science fiction and comics artist, created the cover art for *Das Marsrätsel (The Martian Riddle)* (1957), *Einer von Dreihundert (One of Three Hundred)* (1959), *Der Geist der Legion (The Spirit of the Legion)* (1960), *Der Mars-Robinson* (1964), *Die Marsrebellen (The Mars Rebels)* (1964), *Zwischen den Planeten,* I. Teil, II. Teil *(Between the Planets,* Volumes 1, 2) (1967), *Die Tramps von Luna (The Tramps of Luna)* (1970), *Auf zwei Planeten (On Two Planets)* (1972), *Die Reise zum Mars (The Trip to Mars)* (1973), *Das Marsabenteuer (The Mars Adventure)* (1975), *Der rote Planet (The Red Planet)* (1975), *Jenseits des schweigenden Sterns (Beyond the Silent Star)* (1976).

On this sol in 175 (1939 January 07):

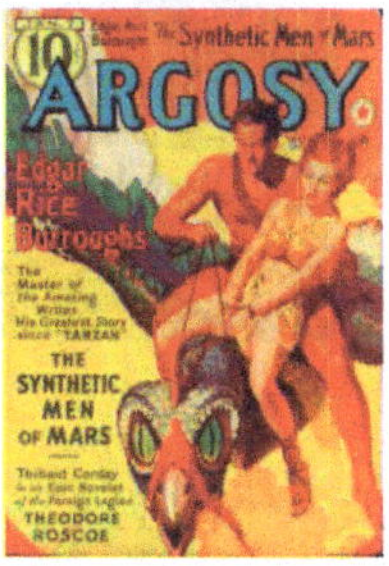

The Synthetic Men of Mars by Edgar Rice Burroughs began serialization in *Argosy.* John Carter desperately needed the aid of Barsoom's greatest scientist. But Ras Thavas was the prisoner of a nightmare army of his own creation, half-humans who lived only for conquest. And in their hidden laboratory seethed a horror that could engulf all Mars.

—Edgarriceburroughs.com

Cover art by Rudolph Belarski.

On this sol in 184 (1955 December 12):

Journey into Space: The World in Peril, Episode 12 airdate. Jet, Lemmy and Doc ignore the Martian's warnings to turn back, and Jet is knocked out after touching a door at the end of the corridor. The crew persuades Paddy to set the ship on course for Mars, so they can warn Earth of the invasion. However, he loses control, and is injured.

—Wikipedia

On this sol in 184 (1987 December 01):

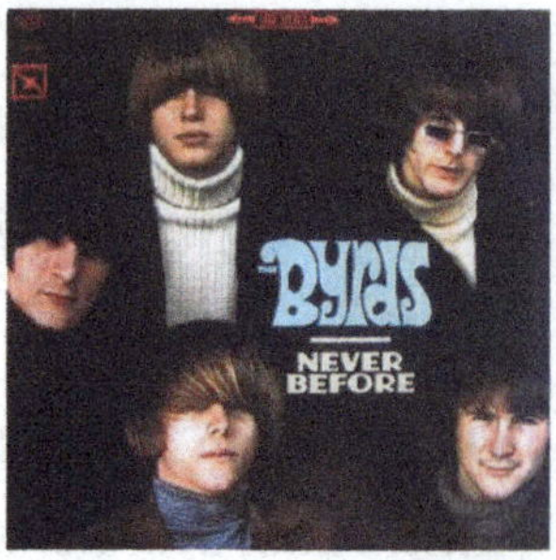

David Crosby's song "Triad" was released on The Byrds' vinyl LP archival album *Never Before* (see 221 Capricornus 12 for lyrics and 221 Mithuna 18 for the song's history). The American rock band recorded the song in 1967 but it was kept off the album *The Notorious Byrd Brothers* (apparently for being too notorious), exacerbating the rift between Crosby and other members of the group.

221 Aries 17 **2025 July 15 10:16 UTC**

Ls 111.0° — Curiosity Sol 4600

MSD 53869 — Perseverance Sol 1575

Earth departure opportunities to Mars via a Type III trajectory and an EEM trajectory.

Mars arrival opportunities from Earth via a Type I trajectory, a Type III trajectory, and an EEM trajectory.

Earth arrival opportunity from Mars via a Type IV trajectory.

On this sol in 171 (1931 July 01):

"The Doom from Planet 4" by Jack Williamson appeared in *Astounding Stories* (date approximate). A ray of fire, green, mysterious, stabs through the night to Dan on his ship. It leads him to an island of unearthly peril.

Interior art by H.W. Wesso.

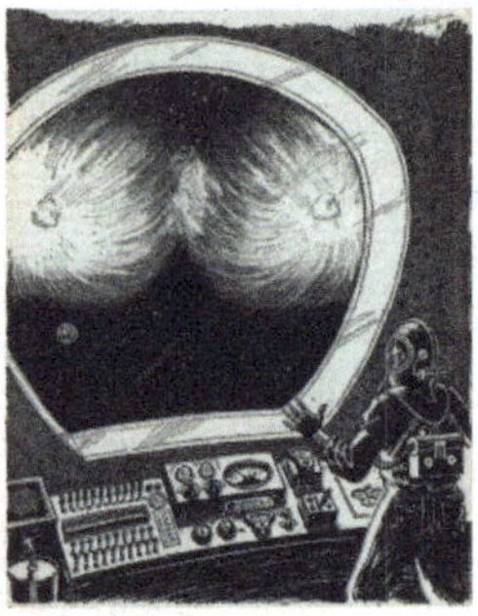

"A Daring Trip to Mars" by Max Valier appeared in *Wonder Stories* (date approximate). Into the comet's deadly bosom they plunged… a hundred-to-one chance to escape….

Interior art by Marco Marchioni.

221 Aries 18 — **2025 July 16 10:56 UTC**

Ls 111.5° — Curiosity Sol 4601

MSD 53870 — Perseverance Sol 1576

As seen from Mars, the Sun enters the constellation Aries.

On this sol in 147 (1886 May 10):

Olaf Stapledon was born; British philosopher; author of *Last and First Men* (1930).

221 Aries 19 — **2025 July 17 11:35 UTC**

Ls 112.0° — Curiosity Sol 4602

MSD 53871 — Perseverance Sol 1577

On this sol in 178 (1944 September 01):

Shadow Over Mars by Leigh Brackett appeared in *Planet Stories* (date approximate). Rick Urquhart was going to conquer the turmoil-ridden planet of Mars. He was penniless and unknown, but there could be no doubt that he would rule the Red Planet-the ancient Martian mystic had made the prophecy, there was no way fate could cheat him of his prize. But there were powerful interests on both Earth and Mars who didn't believe in prophecies-and they were going to undo Rick's future before it had a chance to begin....

—Goodreads.com

Cover art by Earle K. Bergey. Interior art uncredited.

221 Aries 20 — **2025 July 18 12:15 UTC**

Ls 112.4° — Curiosity Sol 4063

MSD 53872 — Perseverance Sol 1578

On this sol in 154 (1899 July 13):

Ronald Davidson was born; American screenwriter, producer, and actor; wrote the screenplays for the motion pictures *Flying Disc Man from Mars* (1950) and *Commando Cody: Sky Marshal of the Universe,* Chapter 7: "Robot Monster from Mars" (1955).

On this sol in 210 (2004 November 08):

MER-A Sol 303: Between Sols 299-303, *Spirit* finished its investigation of Lutefisk. Lutefisk, a rock with some interesting nodules, lies a site roughly 40 meters (131 feet) above and 2700 meters (1.67 miles) away from *Spirit's* landing site on the Gusev plain. Team members should know more about the chemistry of Lutefisk and its nodules when they receive results from the alpha particle X-ray spectrometer and Mössbauer spectrometer.

—NASA/JPL

221 Aries 21 **2025 July 19 12:55 UTC**

Ls 112.9° Curiosity Sol 4064

MSD 53873 Perseverance Sol 1579

On this sol in 106 (1809 April 01):

Honoré Flaugergues drew a sketch of Mars (date approximate).

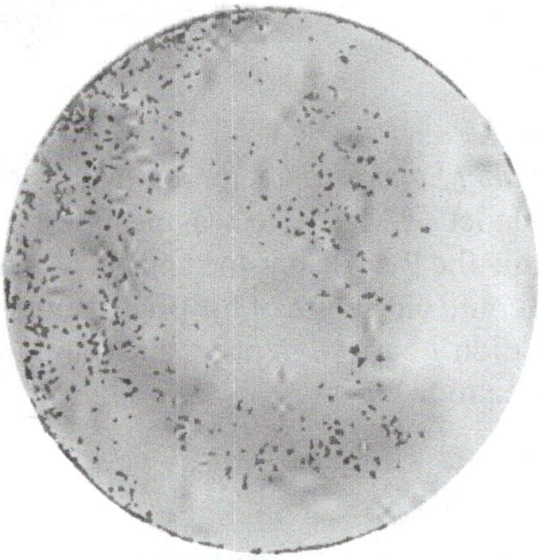

Flaugergues's drawing

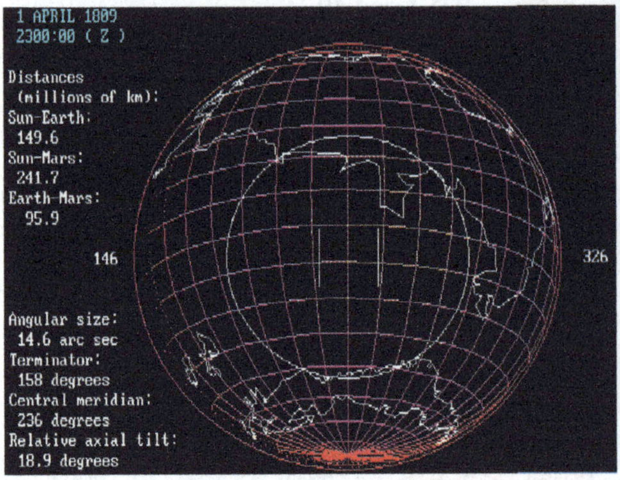

Mars's aspect, 23:00 Z

221 Aries 22 **2025 July 20 13:34 UTC**

Ls 113.4° Curiosity Sol 4065

MSD 53874 Perseverance Sol 1580

On this sol in 210 (2004 November 10):

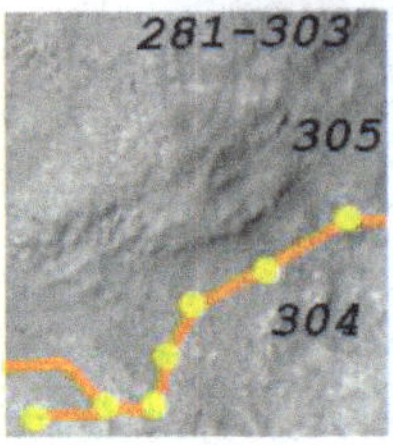

MER-A Sol 305: With the Sun still relatively low on the horizon in the early spring season on Mars, rover drivers are forced to seek driving routes that keep *Spirit* and its solar panels tilted northward for energy reasons. That constraint, plus the rocky terrain, will challenge rover drivers in the coming weeks.

Over the last few weeks, the electrical brakes on *Spirit's* right-front and left-rear steering actuators (motors) apparently failed to disengage during drive attempts. The most likely cause of this anomaly is the buildup of insulating material on the electronic relay contacts that indicate that the brakes are disengaged. To help ensure successful future drives, engineers decided to permanently ignore the "brake-disengaged" indicator. If their theory is correct, the brake will actually be disengaged despite the "failure-to-disengage" indication. If they are wrong, a fuse in the brake circuit will safely blow when they attempt to move the steering actuators. In either case, driving operations will not be adversely affected.

A few sols ago, *Spirit's* engineering team discovered an electric-circuit grounding problem between the rover chassis and the power bus return. This incident occurred at the exact time the *Spirit* team was performing an inspection of the instrument deployment device, or robotic arm. The inspection sequence commanded one of the arm joints to a position beyond where it had previously been. That particular joint, joint number 5, is the rover arm turret, which rotates the four rover arm instruments into position. This coincidence may indicate that the joint 5 move somehow created the electrical short; it could also just be coincidence. The mechanical team has not found any reason to suspect a failure in the joint 5 cabling. To be safe, the engineering team has constrained the use of joint 5 on *Spirit* and *Opportunity* to avoid this extreme position. The constraint is not expected to significantly impact normal operations. The apparent short may also be the result of a failed measurement circuit. The short, if real, has no immediate effect on the rover, but does remove one layer of protection against effects of future shorts should they occur.

—NASA/JPL

221 Aries 23 **2025 July 21 14:14 UTC**

Ls 113.8° Curiosity Sol 4066

MSD 53875 Perseverance Sol 1581

On this sol in 151 (1893 November 23):

H. Thompson Rich was born; American science fiction writer; author of *The Flying City* (1930).

On this sol in 153 (1897 August 28):

Morris Winslow Ankrum was born; American attorney, economics professor, and actor; starred in the motion pictures *Rocketship X-M* (1951), *Red Planet Mars* (1952), and *Invaders from Mars* (1953).

On this sol in 159 (1908 December 10):

MEASURE WATER ON MARS

Percival Lowell Now Says Spectrum Analysis Will Reveal Amount.

Special to *The New York Times.*

BOSTON. Mass., Dec. 9.—Prof. Percival Lowell made to-day a statement with reference to the presence of water vapor on Mars. His assistants, he says, have been able to determine that it is there present and measurable. Precisely how much is still not determined, but it is an amount sufficient to betray Itself in lines of the spectrum, lines which show that the atmosphere of Mars affects the spectrum in the same way. though only one-fifth as much as does the earth's envelope. No one tries to establish for the Martians an air nearly so dense as that which we now breathe. Even the most sanguine of arguers for the planet as a habitation realizes that the smaller body would be in the beginning provided with less of atmosphere-making elements, and that this smaller quantity of air would spread over a surface relatively larger. So it is conceded at the outset that the atmosphere on Mars must be tenuous.

—*The New York Times,* page 4

On this sol in 175 (1939 January 14):

Part 2 of *The Synthetic Men of Mars* by Edgar Rice Burroughs appeared in *Argosy* (date approximate).

On this sol in 176 (1940 December 01):

"Martian Caravan" by Nelson S. Bond appeared in *Blue Book* (date approximate).

Cover art by Herbert Morton Stoops.

"A Product of Mars" by K. Noel Dwyer appeared in *Austra-Fantasy* (date approximate).

On this sol in 184 (1955 December 19):

Journey into Space: The World in Peril, Episode 13 airdate. The asteroid is in chaos and the crew discover Asteroid 786/738 is on its way to sort it out. Mitch and Lemmy go to fetch a bed for Paddy, but become lost and enter the Martian's lair. They hear a familiar voice but can't place it. Paddy dies, and Mitch and Lemmy discover that Jack Evans is really the Lunar Controller, who has been in league with the Martians from the start.

—Wikipedia

On this sol in 210 (2004 November 11):

MER-B Sol 285: The plan was for *Opportunity* to drive 1.5 meters (about 5 feet) east on firm rocky terrain ahead of the rover. The drive went as planned, covering 1.55 meters (5.1 feet). After integrating the results of this drive with an earlier study of Burns Cliff traversability, the team decided not to proceed farther. *Opportunity* has reached the easternmost point of its drive inside Endurance Crater. The rover is at the western edge of Burns Cliff and from this vantage point, it will perform super-high-resolution imaging and other science observations.

—NASA/JPL

221 Aries 24 **2025 July 22 14:53 UTC**

Ls 114.3° Curiosity Sol 4067

MSD 53876 Perseverance Sol 1582

On this sol in 210 (2004 November 12):

MER-B Sol 286: This was a restricted sol for *Opportunity* because the team did not know results of the Sol 285 (210 Aries 23) drive in time for planning this sol. *Opportunity* recorded more than three hours of observations, took a nap, and then used afternoon and overnight communication sessions with *Mars Odyssey.* Solar exposure is excellent inside the crater, so *Opportunity's* power and battery state of charge continue to increase. The rover has not used deep-sleep mode in more than a week, and probably won't for the foreseeable future.

—NASA/JPL

221 Aries 25 **2025 July 23 15:33 UTC**
Ls 114.8° Curiosity Sol 4068
MSD 53877 Perseverance Sol 1583

On this sol in 210 (2004 November 13):

MER-B Sol 287: Sols 287 and 288 were planned together. *Opportunity* began super-high resolution imaging activities on Sol 287. Starting at 11:15 local solar time, the rover performed the following activities: an hour of panoramic camera imaging, an hour of miniature thermal emissions spectrometer imaging and another hour of panoramic camera imaging. Sol 288 was almost exactly the same three-hour activity, but with the images targeted differently.

—NASA/JPL

221 Aries 26 **2025 July 24 16:13 UTC**
Ls 115.3° Curiosity Sol 4069
MSD 53878 Perseverance Sol 1584

On this sol in 210 (2004 November 14):

MER-B Sol 288: The Deep Space Network experienced a station transmitter problem on Sol 287 (210 Aries 25) and *Opportunity* did not receive all of its two-sol uplink as planned. The rover received all except the last part of the Sol 287 bundle, but none of the Sol 288 bundle or data management bundle. Due to quick reaction by the weekend uplink team, bundles were successfully uplinked on Sunday, in time for execution of the Sol 288 plan. The total effect of the missed Sol 287 uplink was a loss of about 30 minutes of science on the morning of Sol 288.

—NASA/JPL

221 Aries 27 **2025 July 25 16:52 UTC**
Ls 115.8° Curiosity Sol 4070
MSD 53879 Perseverance Sol 1585

On this sol in 151 (1893 November 27):

Harry Foster Welch was born; American actor; provided voices for the animated short film "Rocket to Mars" (1946).

On this sol in 154 (1899 July 20):

Laurence Edward Manning was born; Canadian science fiction writer; author of *The Wreck of the Asteroid* (1933).

On this sol in 210 (2004 November 15):

MER-B Sol 289: *Opportunity* Sols 289, 290 and 291 were very similar. Each was a continuation of the remote sensing campaign, with an additional panoramic camera observation. Sol 289 activities included observations of dunes and dust with the panoramic camera and miniature thermal emissions spectrometer. Also, the panoramic camera was used for super-resolution imaging of "Whatanga," a contact boundary between two layers of rocks.

MER-A Sol 310: *Spirit's* panorama looks back to the west.

—NASA/JPL

221 Aries 28 **2025 July 26 17:32 UTC**

Ls 116.2° Curiosity Sol 4071

MSD 53880 Perseverance Sol 1586

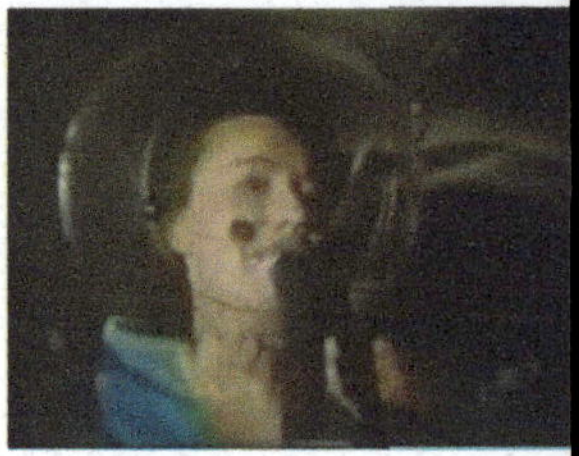

On this sol in 184 (1955 December 24):

Science Fiction Theatre, Season 1 Episode 35, "Project 44." A break through in rocket design "Project 44" means a team of 8 astronauts are to be selected to go on a mission to Mars. Those planning the project have to deal with the emotional stress of such a voyage along with the obvious physical needs.

—International Movie Database

On this sol in 210 (2004 November 16):

MER-B Sol 290: *Opportunity* Sols 289, 290 and 291 were very similar. Each was a continuation of the remote sensing campaign, with an additional panoramic camera observation. In addition to the panoramic camera observation, *Opportunity* made several long-dwell observations of Burns Cliff targets with its miniature thermal emission spectrometer. Cloud observations on the morning of Sol 290 produced a dramatic image.

MER-A Sol 311: *Spirit's* route map through the West Spur of the Columbia Hills.

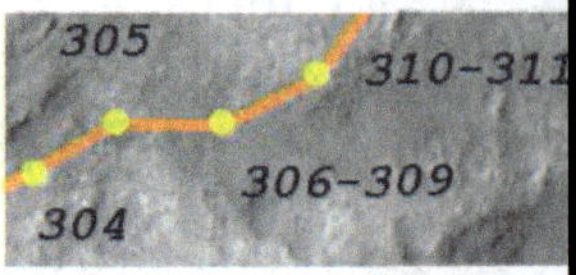

—NASA/JPL

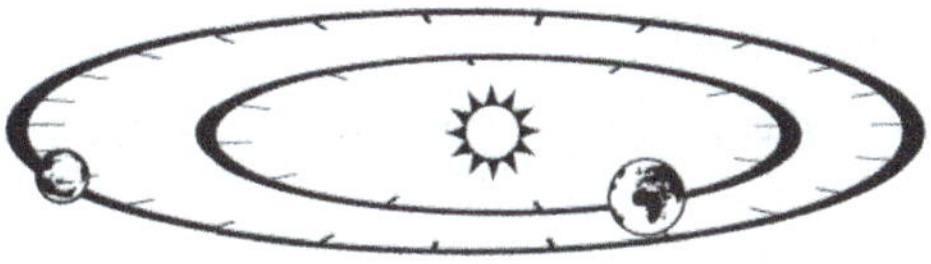

221 MESHA

As seen from Mars, the Sun traverses the central sector of the constellation Aries. The name Mesha (मेष), the Sanskrit word for "ram," is taken from a month in the Hindu solar calendar that originally occurred when the Sun passed through the constellation Aries as seen from Earth.

The median one-way communications delay during this month is 18 minutes 3 seconds.

221 Mesha 01 **2025 July 27 18:11 UTC**

Ls 116.7° Curiosity Sol 4612

MSD 53881 Perseverance Sol 1587

The first sol of the third quarter of Summer on the Aitken calendar; approximately the first sol of the unnamed month between January and February on the Heinlein calendar.

On this sol in 149 (1890 February 23):

Harlan Tarbell was born; American director and actor; directed and starred in the short film "Buck Rogers in the 25th Century" (1934).

On this sol in 210 (2004 November 17):

MER-B Sol 291: *Opportunity* Sols 289, 290 and 291 were very similar. Each was a continuation of the remote sensing campaign, with an additional panoramic camera observation. Sol 291 included a super-resolution observation by *Opportunity* of a target called Bartlett.

The remote sensing campaign is generating a large volume of data at a time when, due to the rover's orientation, there is limited bandwidth available for downlink. As a consequence, *Opportunity* is operating with limited memory headroom, though still within planning guidelines. In order to improve the situation, the team took advantage of the Deep Space Network's 70-meter antenna availability and *Opportunity's* good energy state to plan a one-hour, direct-to-earth session in the middle of the day on Sol 291. This resulted in the downlink of an extra 15 megabits of data.

—NASA/JPL

221 Mesha 02 **2025 July 28 18:51 UTC**

Ls 117.2° Curiosity Sol 4613

MSD 53882 Perseverance Sol 1588

On this sol in 151 (1893 November 30):

Edward Everett Evans was born; American science fiction writer; author of *The Martian and the Vampire* (1950), "Little Miss Ignorance" (1950), "Little Miss Martian" (1951), and "Little Miss Boss" (1952).

On this sol in 175 (1939 January 21):

Part 3 of *The Synthetic Men of Mars* by Edgar Rice Burroughs appeared in *Argosy.*

On this sol in 184 (1955 December 26):

Journey into Space: The World in Peril, Episode 14 airdate. The Lunar Controller tells Mitch and Lemmy that the invasion will be facilitated by a hypnotic television broadcast. Jet and Doc go on a search for the others, and follow conditioned crew members onto the asteroid surface. Wearing Martian suits, they investigate a large sphere, but become stranded on the surface.

—Wikipedia

On this sol in 210 (2004 November 18):

MER-B Sol 292: Having made its closest approach to Burns Cliff, *Opportunity* continued its remote science campaign. In the morning and early afternoon, the rover captured a portion of a color panorama plus images of targets called Cushion and Bartlett. During these observations, *Opportunity* applied some of its excess energy to engage in an hour-long direct-to-earth communications session, downlinking an extra 10 megabits or so. Waking up the next morning at 7:18 local solar time, the rover heated cameras and actuators in advance of looking for clouds.

—NASA/JPL

221 Mesha 03 **2025 July 29 19:31 UTC**

Ls 117.7° Curiosity Sol 4614
MSD 53883 Perseverance Sol 1589

On this sol in 210 (2004 November 19):

MER-B Sol 293: Sols 293 through 295 (210 Mesha 03 through 05) were planned together as a three-sol plan for *Opportunity,* continuing the panorama. Other observations included navigation camera imaging to fill holes in prior image coverage, panoramic camera atmospheric imaging while the Sun was high in the sky, and miniature thermal emission spectrometer imaging of targets Ebony, Ivory, and Cushion. Another direct-to-earth communications session was scheduled for midday. On Sols 293 and 294, the rover finished the remote science campaign, completing the panorama.

—NASA/JPL

221 Mesha 04 **2025 July 30 20:10 UTC**

Ls 118.2° Curiosity Sol 4615
MSD 53884 Perseverance Sol 1590

On this sol in 151 (1893 December 02):

Arthur Julian Zellner was born; American screenwriter and film producer; contributing writer to the motion picture *A Message from Mars* (1921).

On this sol in 159 (1908 December 19):

THE PROPHECY OF MARS

Percival Lowell Interprets the Meaning of the Canals and Predicts the Time When the Earth Will Die of Thirst.

THE NEW SCIENCE OF PLANETOLOGY.

Link Between the Nebular Hypothesis and the Darwinian Origin of Species.

MR. PERCIVAL LOWELL, who has been watching the planet Mars from his high-placed vantage point at Flagstaff, Ariz, and startling the world with fragmentary announcements of how much he has seen there—how infinitely much more inferred—now gives the public the benefit of the total result of. his observation and inference in a volume of less than 300 pages entitled "Mars as the Abode of Life," (The Macmillan Company.) It is only fair to say at once that Mr. Lowell's book at once stirs the imagination and clears the subject marvelously of the clouds of misapprehension with which hasty discussion of his sensational seeming announcements has befogged it.

Mr. Lowell has convinced himself that life in a highly developed state exists upon our neighbor in space, and he has marshaled his material so that it is apt to carry his own conviction over to the reader. But his field is wider, and Mars and the Earth itself serve as illustrations of stages in the vast scheme which Mr. Lowell calls "Planetology "—the science of the making, growth, and death of worlds, by which he links in one complete evolutionary chain the nebular hypothesis at one end and the Darwinian theory of the procession of life from the earliest organic beginnings at the other.

Mr. Lowell's observations of, the disk of the planet [Mars] have convinced him that /the greater portion of Its surface is in a condition resembling that of the Arizona deserts. Indeed, these deserts, seen from a height, have almost, the ochre tint which gives Mars its characteristic ruddy color. Those darker spaces on the planet's surface, which were called seas by the first astronomers who observed them, he finds change hue with the seasons. He argues, therefore, that they are not seas, rather ancient seabeds in which vegetation lingers, since vegetation is the only thing which thus transforms the landscape with the seasons.

The white spots at the two poles of the planet, Mr. Lowell is convinced, are snow fields, because, they, too, wax and wane with the seasons, quite as the polar icecaps of the Earth wax and wane. They are not formed of frozen carbonic acid gas, he says, because when a cap begins to disappear there can be seen at the edge of it toward the Equator a bluish band, which indicates a liquid. Carbonic acid gas does not pass through a liquid stage. As to the surface of the planet careful measurements of the shadows cast at times when the sunrise line stretches across the disk, show no considerable elevations. In other words, the surface is almost level. This should, by the cosmic theory, be the case, for a smaller body, in cooling, having less heat energy, should wrinkle less than a larger body. The fact that the Moon, which is a smaller body even than Mars, possesses mountains as high or higher than any on the Earth, is explained by the fact that the Moon spent the earlier stages of its cosmic life as a part of the Earth, and when it broke off the initial heat of the smaller body was not its own, but that of the Earth.

As already remarked, the greater portion of Mars's surface is utterly arid. The air is extremely thin. The Sun, therefore, shines upon the Equator with a fearful heat. As in the case of the Earth, Mr. Lowell argues, there must arise in equatorial regions a current of heated air which flows north and south. With no mountains to stop it, it holds what moisture it may possess until it reaches the poles, where such moisture is precipitated. In other words, the total precipitation on Mars is snowfall, and all the snow falls at the poles. What the air currents do is to convey each year all the moisture from one pole to the other and back again, leaving the intermediate portions of the planet desert, because water which falls as water evaporates at once. Which brings Mr. Lowell to his canals.

These canals, observed by Schiaparelli and carefully mapped by him many years ago, have been fruitful of doubt and provocative of discussion. They have often been said to exist only in the Imagination or the eye of the observer. But Mr. Lowell at Flagstaff claims not only to have mapped, but to have" photographed them.

Mr. Lowell and his assistants find the- whole surface of the planet, light areas as well as dark, covered by a network of lines. They find that the lines which they call canals, proceed as straight lines or arcs of great circles from point to point. That they are connected each with each. That there are often dark disks, called oases, at the junctions. That the system Is connected with the polar Icecap north and south. The Flagstaff people have further observed that at certain' times this network of geometrical lines is more conspicuous than at others. The lines wax and wane. They have tabulated the figures of visibility obtained from many observations and made elaborate calculations which have convinced them that the lines actually grow wider, and this progressively from the polar cap which has just melted toward the opposite pole. They infer that this indicates the flow of the water through the system from a source to the extremes. Again the lines grow distinct with greater rapidity than they afterward grow indistinct. Mr. Lowell infers, therefore, that the water is admitted at a certain time into the system, but that the effects produced by it pass off gradually. It is impossible, of course, to suppose that what is seen across the vast of space is the actual flowing water. What is seen, Mr. Lowell is sure, is the vegetation which grows up along the line of the canal as the lifegiving water reaches point after point.

Having convinced himself of the existence of a vast system of waterways made for a definite purpose, having convinced himself also that Mars possesses air and water and heat, not much of any three, but enough, if carefully husbanded, to sustain life; having observed that as the conditions of the life on the Earth grow more difficult the creatures of the Earth have grown more ingenious in meeting them, Mr. Lowell naturally assumes the existence of Martians capable of building this system of waterways; of pumping the water through it if necessary.

He goes further. He is bold enough to say that as the Earth grows old and cold, as it loses first its oceans .arid then the vapor of water, as it must, man must also face' the same problem which the Martians-have .solved; that we, also, by our remote descendants, must husband our last supply of water in some such way in order to put off for a few thousand years or so the ultimate extinction of all life on earth, which will come with the end of the water supply.

—*The New York Times,* pages 17-18

On this sol in 210 (2004 November 20):

MER-B Sol 294: Sols 293 through 295 (210 Mesha 03 through 05) were planned together as a three-sol plan for *Opportunity,* continuing the panorama; however, on Sol 294, the rover finished the remote science campaign, completing the panorama.

NASA's Mars Exploration Rover *Opportunity* captured this view of Burns Cliff after driving right to the base of this southeastern portion of the inner wall of Endurance Crater. The view combines frames taken by *Opportunity's* panoramic camera between Sols 287 and 294 (210 Aries 25 and 210 Mesha 04).

This is a composite of 46 different images, each acquired in seven different Pancam filters. It is an approximately true-color rendering generated from the panoramic camera's 750-nanometer, 530-nanometer and 430-nanometer filters. The mosaic spans more than 180 degrees side to side. Because of this wide-angle view, the cliff walls appear to bulge out toward the camera. In reality the walls form a gently curving, continuous surface.

—NASA/JPL

221 Mesha 05 — 2025 July 31 20:50 UTC

Ls 118.6° — Curiosity Sol 4616

MSD 53885 — Perseverance Sol 1591

Earth departure opportunity to Mars via a Type III trajectory.

Mars arrival opportunity from Earth via a Type II trajectory.

Mars arrival opportunity from Earth via an EEM trajectory.

Mars departure opportunity to Earth via a Type III trajectory.

On this sol in 210 (2004 November 21):

MER-B Sol 295: For the first time in 10 sols, *Opportunity* was on the move, beginning a trek out of Endurance Crater. The planned traverse had two parts, with the second leg to be driven only if the rover was on track after the first leg. *Opportunity* drove 3.6 meters (about 12 feet), determined that it was sufficiently close to a designated waypoint, and then continued for another 3.6 meters (about 12 feet). At the completion of the drive, the rover updated information about its attitude and position.

—NASA/JPL

221 Mesha 06 **2025 August 01 21:29 UTC**

Ls 119.1° Curiosity Sol 4617

MSD 53886 Perseverance Sol 1592

Predicted meteor shower associated with parent body C/2007 H2 Skiff: Martian right ascension 319.5°, declination -35.7°. Impacting the daytime hemisphere south of the equator.

On this sol in 210 (2004 November 22):

MER-B Sol 296: *Opportunity* continued its westward journey, performing an 11-meter (36-foot) drive on flat rocky terrain that provides good traction for the rover. The drive succeeded as planned, leaving the vehicle perfectly in the middle of its intended path. *Opportunity* then performed two hours of post-drive observations. At this point, there appeared to be a possible shortcut out of the crater, just about 13 meters (about 43 feet) in front of *Opportunity.* If traversable, this early egress chute could shorten the vehicle's exit route by more than 30 meters (about 98 feet) and many sols. In response to this possibility, the team planned further investigation of the chute area.

—NASA/JPL

221 Mesha 07 **2025 August 02 22:09 UTC**

Ls 119.6° Curiosity Sol 4618

MSD 53887 Perseverance Sol 1593

On this sol in 210 (2004 November 23):

MER-B Sol 297: Sols 297 through 299 (210 Mesha 07 through 09) and Sols 300 through 302 (210 Mesha 10 through 12) were planned as two consecutive three-sol plans for *Opportunity* due to the Thanksgiving holiday on Earth. Commands for Sols 297, 298 and 299 were uplinked on Sol 297 and commands for Sols 300 through 302 (210 Mesha 10 through 12) were uplinked on Sol 298.

The plan for Sol 297 included a drive west for about 13 meters (43 feet) to a relativity flat area. This location would afford a good view of the possible egress chute and allow the robotic arm to be deployed so the Mossbauer spectrometer could be placed on the filter magnet and perform four sols of data collection. The drive stopped early because *Opportunity* slipped more than anticipated and missed an intermediate waypoint. The drive covered 8.6 meters (about 28 feet). However, imaging after the drive allowed a detailed analysis of the possible shortcut.

To make an early exit, *Opportunity* would have had to cross terrain sloping 28 degrees. The opening of the chute is 1.14 meters (about 4 feet) wide, and there appears to be tall rock outcropping very close to the opening. On the route to the originally planned exit path at Keratepe, where *Opportunity* entered the crater six months ago, the average slope is only 22 degrees and there are no large obstacles to avoid. So the decision was made to skip the shortcut and continue toward Karatepe.

—NASA/JPL

221 Mesha 08 **2025 August 03 22:48 UTC**
Ls 120.1° Curiosity Sol 4619
MSD 53888 Perseverance Sol 1594

On this sol in 183 (1954 February 13):

Rod Brown of the Rocket Rangers Season 1 Episode 44 "Return of the Stickmen" airdate.

—International Movie Database

On this sol in 184 (1956 January 02):

Journey into Space: The World in Peril, Episode 15 airdate. While the Lunar Controller is away, Jet and Doc enter his quarters from the asteroid surface. Mitch and Lemmy explain what has happened to them, and they hatch a plan to imprison the Lunar Controller and escape.

—Wikipedia

On this sol in 210 (2004 November 24):

MER-B Sol 298: *Opportunity* operated autonomously during the second sol of the first three-sol plan for to the Thanksgiving holiday on Earth.

—NASA/JPL

221 Mesha 09 **2025 August 04 23:28 UTC**
Ls 120.6° Curiosity Sol 4620
MSD 53889 Perseverance Sol 1595

On this sol in 146 (1884 July 12):

Edgar Stehli was born; French-American actor; starred in the *Tales of Tomorrow* episode "The Crystal Egg" (1951).

On this sol in 175 (1939 January 28):

Part 4 of *The Synthetic Men of Mars* by Edgar Rice Burroughs appeared in *Argosy*.

On this sol in 210 (2004 November 25):

MER-B Sol 299: *Opportunity* operated autonomously during the last sol of the first three-sol plan for to the Thanksgiving holiday on Earth.

—NASA/JPL

221 Mesha 10 **2025 August 06 00:08 UTC**
Ls 121.0° Curiosity Sol 4621
MSD 53890 Perseverance Sol 1596

On this sol in 210 (2004 November 26):

MER-B Sol 300: *Opportunity* operated autonomously during the first sol of the second three-sol plan for to the Thanksgiving holiday on Earth.

—NASA/JPL

221 Mesha 11 **2025 August 07 00:47 UTC**

Ls 121.5° Curiosity Sol 4622
MSD 53891 Perseverance Sol 1597

On this sol in 151 (1893 December 10):

Luis Yakovich Braunshtejn (Луис Якович Браун-штейн, a.k.a. Lew Brown) was born; Russian-American songwriter, composer, author, publisher and film producer; wrote the story and dialogue for *Just Imagine* (1930) as well as lyrics for the songs "(There's Something About) An Old-Fashioned Girl," "Mothers Ought to Tell Their Daughters," "(I Am the Words) You Are the Melody," "The Drinking Song," "Never Swat a Fly," and "The Romance of Elmer Stremingway."

On this sol in 152 (1895 October 28):

THE LATEST LIGHT ON MARS.

MARS. By Percival Lowell. 8vo. With illustrations. New-York: Houghton, Mifflin & Co.

The great question concerning Mars is whether or not the planet is inhabited. It was in the hope of obtaining data for an intelligent opinion on this question that the observations were made which are described in this book, a reprint of the author's recent papers in *The Atlantic*. Prof. Lowell systematically studied the planet from May 24, 1894, to April 3, 1895, in the observatory at Flagstaff, Arizona. Prof. W. H. Pickering and A. E. Douglass were associated with him in the observations.

The conclusion Prof. Lowell draws as to the main question is that it is probable there is some form of life on Mars. More than this he does not undertake to say. He holds out the hope that some day the people of this world may know with certainty whether there is life on Mars, but for the present he says we must deal with probabilities. We also are cautioned against assuming that Mars, if inhabited, must be inhabited by beings of our own kind. Extra-terrestrial life does not mean necessarily extra-terrestrial human life.

The substance of human knowledge as to Mars seems to be that the astronomers have observed on the face of the planet what probably we may regard as indications that intelligent beings were at work there. A glance at a map of Mars is sufficient to suggest that the network of canals which is shown on it is explainable far easier upon the hypothesis of local intelligence than upon the alternative hypothesis that it has been made by natural causes. If the maps the astronomers have given us are trustworthy, they seem to decide the case.

Practically, the entire question whether there are intelligent beings living on Mars must hinge in the layman's mind on the reasonableness of the proposition that it is possible to make an accurate map of something 35,000,000 miles away. The fact that independent observations and studies, conducted at various times and in various parts of the world, result in maps which in all important respects are duplicates, one of another, would seem to destroy all ground for reasonable doubt that Mars presents the appearance the maps indicate. Prof. Lowell turns with disgust from the hypothesis that natural causes have made all the canals and spots on the face of Mars which he has seen through his telescope.

There is no inherent improbability in the proposition that there may be intelligent life on our near neighbor in the planetary world; but, if Mars is like the maps we have of it, it is improbable that nature made it what it is—nature is not given to working with chain and compass.

Prof. Lowell has stated for us briefly the basis of his opinion that there is a Martian people. As he puts it:

> We find, in the first place, that the broad physical conditions of the planet are not antagonistic to some form of life; secondly, that there is an apparent dearth of water upon the planet's surface, and, therefore, if beings of sufficient intelligence inhabited it, they would have to resort to irrigation to support life; thirdly, that there turns out to be a network of markings, covering the disk, precisely counterparting what a system of irrigation would look like; and, lastly, that there is a set of spots placed where we should expect to find the lands thus artificially fertilized, and behaving as such constructed oases should. All this, of course, may be a set of coincidences, signifying nothing; but the probability points the other way.

The Martians may be altogether different from us. Prof. Lowell tells us that if there are Martians, and they are in human form, there is nothing to prevent their being three times as big as we are, and on account of the nature of the Martian atmosphere and the quality of Martian gravity capable of doing fifty times as much work as we are able to do, man for man. This ratio is almost like that between man and a powerful steam shovel. If, then, the Martians are giants, each of them as useful in the work of building canals as one of our steam shovels, the problem is simplified somewhat. How long would this country hesitate over the Nicaragua Canal project if it could hire 10,000 men, each equal in efficiency to a steam shovel, to start in on the great divide? The canal would be built inside of six months.

But there is another consideration, of which Prof. Lowell reminds us:

> Quite possibly such Martian folk are possessed of inventions of which we have not dreamed, and with them electrophones and kinetoscopes are things of a bygone past, preserved with veneration in museums as relics of the clumsy contrivances of the simple childhood of the race. Certainly what we see hints at the existence of beings who are in advance of, not behind us, in the journey of life.

According to this hypothesis, we must think of the Martians as equipped with knowledge how to make canals easily and quickly. They must have power such as we never have dreamed of, apparatus that is gigantic and admirably adapted to canal-making. They must be most astonishing adepts in ditchmaking. Without canals they must die; their great necessity has stimulated their wonderfully developed inventive powers, so that they are confident of their ability to keep themselves watered, or the present at least, providing they shall continue to multiply according to the established rate of increase. Their life depends on the continuance of the ice belts at the poles of their globe. While the ice shall continue to form there and melt they will live and have what they shall need to eat and drink.

But what if the Martians, if there be any, are something else than men? If we talk of probabilities we must admit at the start that the probability is exceedingly strong that the Martians are not at all like us.

Assuming that they have criss-crossed their globe with great irrigating ditches, it is strongly probable that they are different. Prof. Lowell remarks:

> Even on this earth man is in the nature of an accident. He is the survival of by no means the highest physical organism. He is not even a high form of mammal. Mind has been his making. For aught we can see, some lizard or batrachian might just as well have popped out his plan early in the race and been now the dominant creature of this earth. Under different physical conditions he would have been certain to do so. Amid the surroundings that exist on Mars—surroundings so different from our own—we may be practically sure other organisms have been evolved of which we have no cognizance. What manner of beings they may be we lack the data even to conceive.

—*The New York Times,* page 3

On this sol in 156 (1903 May 07):

Martin Brandt was born; German actor; starred on the *Tales of Tomorrow* episode "The Last Man on Earth" (1951).

On this sol in 179 (1946 August 09):

The animated short film "Rocket to Mars" was released. Popeye and Olive are touring a museum when they accidentally launch a rocketship to Mars. Olive escapes, but Popeye gets to Mars, where he is attacked (by a group led by Bluto) that was preparing to invade Earth. Fortunately, Popeye has a can of spinach handy, so he can save the Earth (turning most of the Martian war apparatus into amusement park rides).

—Jon Reeves <jreeves@InternationalMovieDatabase.com>

On this sol in 210 (2004 November 26):

MER-B Sol 301: *Opportunity* operated autonomously during the second sol of the second three-sol plan for to the Thanksgiving holiday on Earth.

—NASA/JPL

221 Mesha 12 **2025 August 08 01:27 UTC**

Ls 122.0° Curiosity Sol 4623

MSD 53892 Perseverance Sol 1598

On this sol in 210 (2004 November 27):

MER-B Sol 302: *Opportunity* operated autonomously during the last sol of the second three-sol plan for the Thanksgiving holiday on Earth.

—NASA/JPL

221 Mesha 13 **2025 August 09 02:06 UTC**

Ls 122.5° Curiosity Sol 4624
MSD 53893 Perseverance Sol 1599

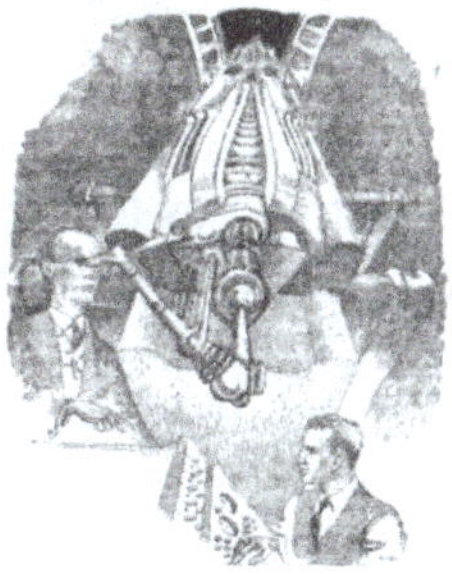

On this sol in 175 (1939 February 01):

"The Message from the Void" by Otto Binder (as Hubert Mavity) appeared in *Dynamic Science Stories* (date approximate). Mutual misunderstanding between Mars and Earth because of two factors in the abstract beyond their comprehension!

Interior art by Jack Binder.

221 Mesha 14 **2025 August 10 02:46 UTC**

Ls 123.0° Curiosity Sol 4625
MSD 53894 Perseverance Sol 1600

On this sol in 210 (2004 December 01):

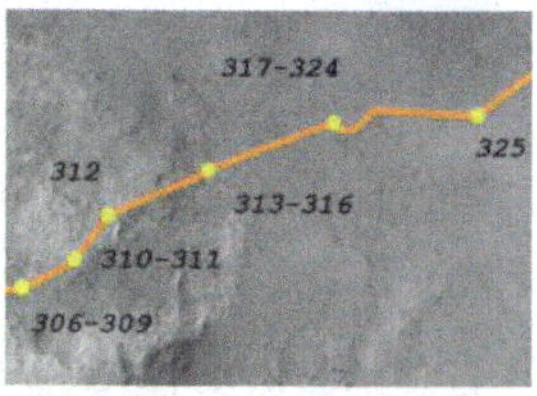

MER-A Sol 325: Between Sols 306 and 325 (210 Mesha 05 and 14), *Spirit* finished shooting a Thanksgiving panorama with the panoramic camera; investigated new rock targets Corn, Cocomama, and Butter with the science instruments; and continued to drive eastward between the West Spur and an area where the terrain slopes back upward toward Husband Hill, named for astronaut Rick Husband, commander of Space Shuttle *Columbia's* ill-fated STS-107 mission.

During the 19 Sols ending on Sol 325, *Spirit* continued to explore in the Columbia Hills. *Spirit* reached a position northeast of a ridge called Machu Picchu and began crossing a 200-meter-wide (656-foot-wide) flat saddle area.

Spirit successfully completed about 80 meters (262 feet) of driving, bringing the total mission traverse to 3.82 kilometers (2.37 miles).

The amount of electric current drawn by the motor of the right front wheel continues to be a concern. However, during a recent drive the current draw was closer to normal than it had been in preceding weeks. Engineers will continue to limit use of this wheel by driving backwards when terrain allows, dragging it 90 percent of the time.

—NASA/JPL

221 Mesha 15 **2025 August 11 03:26 UTC**

Ls 123.5° Curiosity Sol 4626
MSD 53895 Perseverance Sol 1601

On this sol in 184 (1956 January 09):

Journey into Space: The World in Peril, Episode 16 airdate. The crew imprison the Lunar Controller, and gain his assistance by threatening to release him into space. Lemmy smashes up the computerised brain, to stop messages being sent to the rest of the Martian fleet. Frank pilots the ship as they escape the asteroid and set course for Mars, but they find *Discovery* to have been plundered.

—Wikipedia

On this sol in 210 (2004 December 02):

MER-A Sol 326: *Spirit* drove five of the last seven days, continuing its trek towards the top of Husband Hill. *Spirit's* intermediate goal is a ridge dubbed Larry's Lookout, which is roughly 75 meters (246 feet) away. Getting there using *Spirit's* current path will be a challenge given the sand, slope, and rocks in this area. *Spirit* paused for a set of weekend observations of a rock called Wishstone. Total odometry for the mission is now 3,944 meters (2.45 miles).

Spirit's traverse up Husband Hill

The amount of electric current drawn by the motor on the right front wheel is in the normal range. Near the end of *Spirit's* long series of drives from Bonneville Crater to the Columbia Hills, the right front wheel began to draw roughly twice the current of the other five wheels. The increased current prompted engineers to limit the use of this wheel to preserve its life. Since arriving at the hills, *Spirit* has had relatively few driving days. *Opportunity* team's current working theory on this problem suggests that the recent rest periods have allowed the lubricant in this wheel to redistribute, causing the current draw to return to normal. Periodic rest days will be included in rover drivers' plans, and *Spirit* will alternate forward and backward driving to keep the lubricant in all of the wheels more evenly distributed.

—NASA/JPL

221 Mesha 16 — **2025 August 12 04:05 UTC**

Ls 123.9° — Curiosity Sol 4627

MSD 53896 — Perseverance Sol 1602

On this sol in 175 (1939 February 04):

Part 5 of *The Synthetic Men of Mars* by Edgar Rice Burroughs appeared in *Argosy.*

Interior art by Samuel Cahan.

On this sol in 210 (2004 December 02):

MER-B Sol 306: This was the second sol of *Opportunity*

inspecting a rock target called Paikea with tools on the robotic arm. On the previous sol, *Opportunity* had cleaned the surface of Paikea with its rock abrasion tool brush. During Sol 306, the rover observed the target with the panoramic camera and the microscopic imager, then ground away a patch of the rock's surface for about two hours with the rock abrasion tool. After the grind, the rover examined the fresh hole in the rock with the panoramic camera, hazard-avoidance camera and microscopic imager. This was followed by placement of the alpha particle X-ray spectrometer on the abraded area for later data collection.

—NASA/JPL

221 Mesha 17 **2025 August 13 04:45 UTC**

Ls 124.4° Curiosity Sol 4628
MSD 53897 Perseverance Sol 1603

On this sol in 148 (1888 April 24):

Arnold Lorne Hicks was born; Canadian-American pulp artist; created the interior art for "Hide and Seek" by Arthur C. Clarke (1949).

On this sol in 210 (2004 December 03):

MER-B Sol 307: *Opportunity* woke at about 4:45 a.m. (local solar time) for a morning communication-relay pass with *Mars Odyssey.* Then it turned on the alpha particle X-ray spectrometer and gathered data until mid-morning.

Sols 307 through 309 (210 Mesha 17 through 19) were planned together as a three-sol plan. Sol 307 was similar to Sol 306 (210 Mesha 16), with imaging, rock grinding, and overnight X-ray spectrometer measurements, this time targeted on Wharenhui. Also, between 1:30 p.m. and 2 p.m. local solar time, *Opportunity* used its miniature thermal emission spectrometer and panoramic camera to make observations coordinated with a *Mars Global Surveyor* overflight.

—NASA/JPL

On this sol in 210 (2004 December 04):

MER-A Sol 328: Spirit route map through the West Spur of the Columbia Hills.

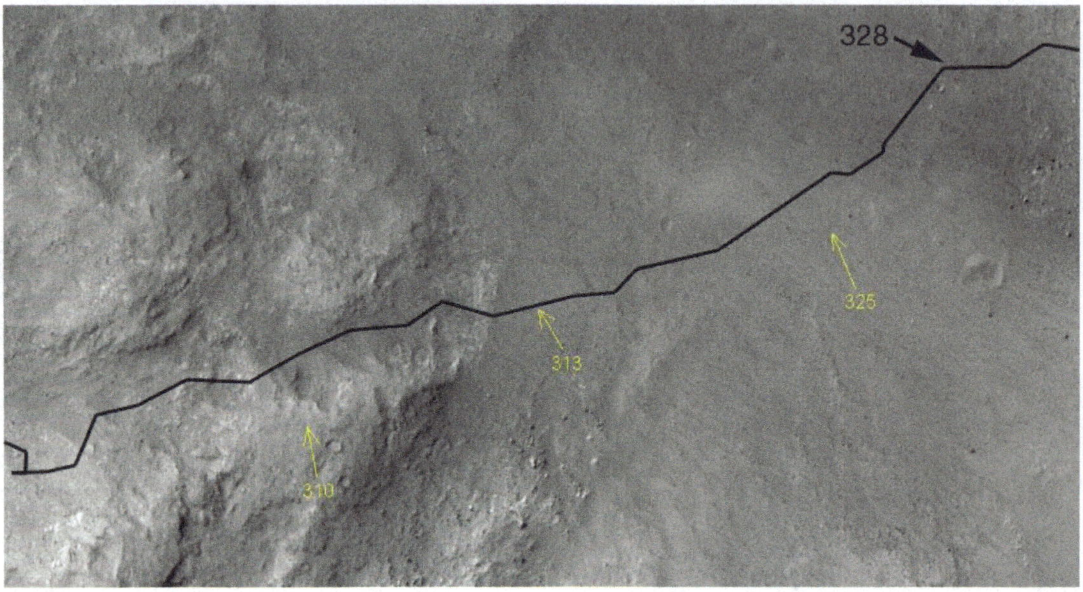

221 Mesha 18 **2025 August 14 05:24 UTC**

Ls 124.9° Curiosity Sol 4629

MSD 53898 Perseverance Sol 1604

On this sol in 186 (1959 October 17):

Augie Doggie and Doggie Daddy, Season 1 Episode 5, "High and Flighty" airdate. Daddy chases after a flying saucer headed for Mars piloted by Augie.

On this sol in 210 (2004 December 04):

MER-B Sol 308: *Opportunity* continued routine atmospheric observations and remote sensing of the heat shield debris field. In the late morning, the rover placed its Mössbauer spectrometer on the freshly drilled hole in Wharenhui, and then collected data nearly continuously for the next two sols.

On this sol in 210 (2004 December 05):

MER-A Sol 329: On Sols 327 to 329 (210 Mesha 16 to 18), the Mössbauer spectrometer was placed on the compositional calibration target for a series of observations over the Earth weekend. This is done periodically to calibrate the Mössbauer instrument.

—NASA/JPL

221 Mesha 19 **2025 August 15 06:04 UTC**

Ls 125.4° Curiosity Sol 4630
MSD 53899 Perseverance Sol 1605

On this sol in 189 (1965 June 08):

Under contract for NASA's Marshall Space Flight Center, General Dynamics/Convair issued its report, "Manned Mars and Venus Exploration Study." The report compared flyby and capture missions and a number of mission models for sending a crew of eight to Mars sometime in the period from 1973 through 1990. For the study's reference mission, departure from Earth orbit was scheduled for 1982 January 17. Launch windows occurring in 1984, 1986, 1988, and 1990 were also considered. It was estimated that, based upon the 1982 departure date, the crew would arrive at Mars on 1982 July 27. Following a month in the vicinity of the Red Planet, the astronauts would depart for Earth on 1982 August 25, and arrive home on 1983 February 24.

On this sol in 210 (2004 December 05):

MER-B Sol 309: In order to provide sufficient energy for extended integration of the Mössbauer spectrometer on the hole drilled in Wharenhui, *Opportunity's* overnight communication pass with *Mars Global Surveyor* was sacrificed.

—NASA/JPL

221 Mesha 20 **2025 August 16 06:44 UTC**

Ls 125.9° Curiosity Sol 4631
MSD 53900 Perseverance Sol 1606

Earth departure opportunity to Mars via an EEM trajectory.

Mars arrival opportunity from Earth via a Type II trajectory.

Mars arrival opportunities from Earth via an EVM trajectory, an EEM trajectory, and a Type IV trajectory.

Earth arrival opportunity from Mars via a Type II trajectory.

On this sol in 150 (1892 February 01):

"The region of the south pole was subject to marked changes of color and brightness. When first observed in July it was described as red, indicating a lack of vegetation. Later it turned yellow, indicating cloud. On one or two occasions it was thought to be white, indicating snow. On February 1 it was described as dark greenish grey, darker than the rest of the terminator...."

—William H. Pickering, 1915

According to the Pickering calendar, the Martian date was July 41.

On this sol in 210 (2004 December 06):

MER-B Sol 310: *Opportunity's* grinding activity was not as productive as hoped, so plans for the next couple of sols were revised. Because the team had to wait for retransmission of some rock abrasion tool data, the sol was spent using the microscopic imager, placing the alpha particle X-ray spectrometer, and performing about an hour of remote-sensing observations.

—NASA/JPL

221 Mesha 21 **2025 August 17 07:23 UTC**
Ls 126.4° Curiosity Sol 4632
MSD 53901 Perseverance Sol 1607

On this sol in 210 (2004 December 07):

MER-B Sol 311: *Opportunity* spent the sol re-grinding Wharenhui. The plan was to grind another 3 to 4 millimeters (0.1 to 0.2 inches) into the rock. The grind went as planned, microscopic images were taken and the alpha particle X-ray spectrometer was placed in the hole. That spectrometer's integration was not started until the Sol 312 early-morning *Mars Odyssey* pass at 4:20 a.m. Mars local time.

On this sol in 210 (2004 December 08):

MER-A Sol 332: *Spirit* drove five of the last seven days, continuing its trek towards the top of Husband Hill. *Spirit's* intermediate goal is a ridge dubbed Larry's Lookout, which is roughly 75 meters (246 feet) away. Getting there using *Spirit's* current path will be a challenge given the sand, slope, and rocks in this area. *Spirit* paused for a set of weekend observations of a rock called Wishstone. Total odometry for the mission is now 3,944 meters (2.45 miles).

The amount of electric current drawn by the motor on the right front wheel is in the normal range. Near the end of *Spirit's* long series of drives from Bonneville Crater to the Columbia Hills, the right front wheel began to draw roughly twice the current of the other five wheels. The increased current prompted engineers to limit the use of this wheel to preserve its life. Since arriving at the hills, *Spirit* has had relatively few driving days. The rover team's current working theory on this problem suggests that the recent rest periods have allowed the lubricant in this wheel to redistribute, causing the current draw to return to normal. Periodic rest days will be included in rover drivers' plans, and *Spirit* will alternate forward and backward driving to keep the lubricant in all of the wheels more evenly distributed.

—NASA/JPL

221 Mesha 22 **2025 August 18 08:03 UTC**
Ls 126.9° Curiosity Sol 4633
MSD 53902 Perseverance Sol 1608

On this sol in 175 (1939 February 11):

Part 6 (the conclusion) of *The Synthetic Men of Mars* by Edgar Rice Burroughs appeared in *Argosy.*

On this sol in 184 (1956 January 16):

Journey into Space: The World in Peril, Episode 17 airdate. They head into orbit, transfer to Freighter #1 and, using a modified radio transmitter, attempt to contact Earth, but they cannot get their warning about the invasion through. They spot the asteroid invasion fleet approaching them from behind. Jet decides to use all remaining fuel to set on course for Earth, in order to increase their chances of warning Earth of the invasion.

—Wikipedia

On this sol in 210 (2004 December 08):

MER-B Sol 312: *Opportunity's* alpha particle X-ray spectrometer integration was not started until the early-morning *Mars Odyssey* pass at 4:20 a.m. Mars local time.

On this sol in 210 (2004 December 09):

MER-A Sol 333: *Spirit* used the rock abrasion tool brush to scrub a small section of Wishstone and took microscopic images of the spot. *Spirit* then placed the alpha particle X-ray spectrometer on the spot for collecting data overnight.

Atmospheric observations using the miniature thermal emission spectrometer, navigation camera, and panoramic camera continue on a daily basis.

—NASA/JPL

221 Mesha 23 **2025 August 19 08:42 UTC**

Ls 127.4° Curiosity Sol 4634
MSD 53903 Perseverance Sol 1609

On this sol in 210 (2004 December 09):

MER-B Sol 313: This was a restricted sol for *Opportunity* because results from the Sol 312 drive were not available for planning Sol 313. That meant that no driving or robotic-arm activities were permitted. So, *Opportunity* performed about two hours of observations using the panoramic camera and miniature thermal emission spectrometer and then went to sleep in the early afternoon. The rover woke up to support late-afternoon and early-morning communication relays by the orbiting *Mars Odyssey*.

On this sol in 210 (2004 December 10):

MER-A Sol 334: *Opportunity* performed 90 minutes of remote-sensing observations, inspected debris from the heat shield's flank with the microscopic imager, and then placed the alpha particle X-ray spectrometer on the capture magnet on *Opportunity's* solar panel.

—NASA/JPL

221 Mesha 24 **2025 August 20 09:22 UTC**

Ls 127.9° Curiosity Sol 4635
MSD 53904 Perseverance Sol 1610

Predicted meteor shower associated with parent body C/1932 G1 Houghton–Ensor: Martian right ascension 310.4°, declination -24.5°.

On this sol in 185 (1957 December 04):

The Magical World of Disney Season 4 Episode 12 "Mars and Beyond" airdate. The episode begins with an introduction of Walt Disney and his robot friend Garco, who provide a brief overview of this episode, which starts with a look at mankind seeking to understand his world, first noticing patterns in the stars. He develops beliefs regarding the celestial bodies. Theories from scientists and philosophers are discussed. Ptolemy's inaccurate but formerly-accepted theories are discussed, as are those of Copernicus. Life on other planets is considered, soon focusing on Mars.

Ideas from science-fiction authors H.G. Wells and Edgar Rice Burroughs are brought to life with colorful animation. Pulp science fiction comics of the time are parodied. Then the program adopts a serious tone as it profiles each of the planets in the solar system, from the perspective of what would happen to man on them. The claim is that whereas most of the planets are either too cold or too hot for life as we know it, life on Mars could almost be normal, something that is of importance for the future. Dr. E.C. Slipher then discusses Mars and the possibility that life already exists there. More animation speculates what the conditions on Mars might be like. This section is filled with striking, inventive and decidedly atypical Disney animation. The show wraps up with what a trip to Mars would entail for a space crew and its vessels. Contributor/spacecraft designer Ernst Stuhlinger presents his design and details regarding a unique umbrella-shaped MarsShip: the top portion is a revolving outer quarters ring providing artificial gravity for the crew of 20, under "parasol" coolant tubes. At the other end is a sodium-potassium reactor to provide power to the midsection electric/ion drive. Attached upright is a chemically-fueled winged tail-lander. The mission shown involves six MarsShips, ultimately reaching 100,000 MPH, taking a 400-day spiral course to Mars where they would spend 412 days on the surface before returning.

—International Movie Database

On this sol in 210 (2004 December 10):

MER-B Sol 314: *Opportunity's* Sols 314 through 316 were planned in another single planning cycle. The plan was to complete the egress from Endurance Crater on Sol 315, so Sol 314 was another remote sensing sol. This would be the last full sol inside Endurance. *Opportunity* spent about two and a half hours observing with the panoramic camera and miniature thermal emission spectrometer. It also performed a nighttime observation with the miniature thermal emission spectrometer just before midnight. To ensure that *Opportunity* had adequate power, the early-morning communication-relay session with *Mars Odyssey* was canceled and *Opportunity* went into a modified deep sleep after completing the late-night observation.

On this sol in 210 (2004 December 11):

MER-A Sol 335: *Spirit* removed the alpha particle X-ray spectrometer from the hole and replaced it with the Mössbauer spectrometer. *Spirit* also started a long series of Mössbauer observations that would last until the early morning of Sol 337 (210 Mesha 26).

—NASA/JPL

221 Mesha 25 **2025 August 21 10:01 UTC**

Ls 128.3° Curiosity Sol 4636

MSD 53905 Perseverance Sol 1611

On this sol in 137 (1867 August 25):

In Chapter 1 of Percy Greg's *Across the Zodiac* (1880), the spacecraft *Astronaut* crashed on an island in the South Pacific. According to his Martialists' reckoning, the date was 275th sol of the marsrev 13238.

On this sol in 156 (1903 May 21):

Manly Wade Wellman was born; American science fiction writer; author of *The Disc-Men of Jupiter* (1931) and "When Planets Clashed" (1931), and co-author of "The Adventure of the Martian Client" (1969), "Sherlock Holmes Versus Mars" (1975), two of the stories in *Sherlock Holmes's War of the Worlds* (1975).

On this sol in 176 (1941 January 01):

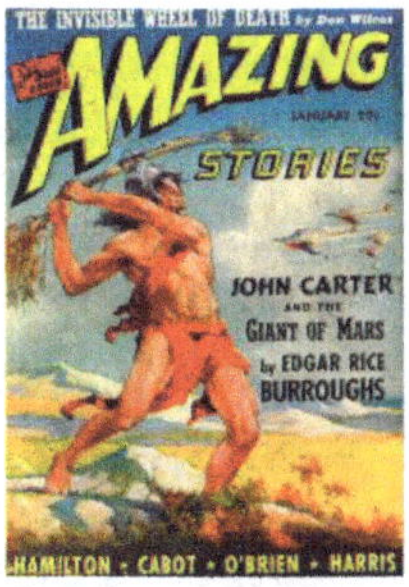

"John Carter and the Giant of Mars" by Edgar Rice Burroughs appeared in *Amazing Stories* (date approximate). Co-written with his son, John Coleman Burroughs, it follows the story of the adventurous Carter and the princess he loves, Dejah Thoris. As the book opens, the two of them are attacked, and Dejah is carried away, leaving Carter to wonder whether he will ever see her again. He calls on his friend Tars Tarkas to help, but their journey to recover the princess is an eventful and sometimes dangerous one. Among other encounters, they have to face a city populated entirely by rats—not to mention, of course, the giant of Mars—who is no less than 130 feet tall!

—Amazon.com

Cover art by J. Allen St. John.

"Under the Sand-Seas" by Oliver E. Saari appeared in *Super Science Stories* (date approximate). Six hundred miles of devouring red desert lay between them and the nearest city—and the bust of a long-dead Martian smiled with the knowledge that could get them there!

Interior art by Leo Morey.

On this sol in 210 (2004 December 11):

MER-B Sol 315: *Opportunity* has finished its work inside Endurance Crater and climbed out. Before leaving, the rover examined a transition point between dark and light rock layers about 20 meters (about 66 feet) from the rim of the crater. Communication with *Mars Odyssey* has been good, so the backlog of onboard data has improved. The rover spent six months inside the stadium-sized crater to study layered bedrock exposed there. The exit drive on Sol 315 put *Opportunity* completely outside the crater for the first time since Sol 134 (210 Makara 11). *Opportunity* continues to be in excellent health.

—NASA/JPL

221 Mesha 26 **2025 August 22 10:41 UTC**

Ls 128.8° Curiosity Sol 4637
MSD 53906 Perseverance Sol 1612

On this sol in 160 (1910 November 29):

Al Schwartz was born; American television writer and director; wrote the teleplay for *Gilligan's Island*, Season 2 Episode 4, "Smile, You're on Mars Camera" (1965).

On this sol in 210 (2004 December 12):

MER-B Sol 316: This was the third sol of a three-sol plan, and because *Opportunity* had driven on Sol 315, Sol 316 was restricted to remote-sensing observations. The rover performed about two hours of remote sensing and went to sleep. Out on the plains, Opportunity went from a northerly tilt that is very good for solar exposure, to a southerly tilt that is not so good for solar exposure. The tilt was expected to be as high as 10 degrees, but Opportunity's actual tilt was about 5 degrees. Daily output from the solar panels went from 840 watt-hours in the crater, to 730 watt-hours on the plains.

On this sol in 210 (2004 December 13):

MER-A Sol 337: *Spirit* stowed its robotic arm, then bumped backwards to take final images of Wishstone and the rock abrasion tool hole. *Spirit* was commanded to drive 15 meters (49 feet), but drove only about 6 meters (20 feet) due to experiencing slippage of up to 80 percent on uphill portions of the drive.

—NASA/JPL

221 Mesha 27 **2025 August 23 11:21 UTC**

Ls 129.3° Curiosity Sol 4638
MSD 53907 Perseverance Sol 1613

On this sol in 210 (2004 December 13):

MER-B Sol 317: Since the *Opportunity* team continues to be operating in restricted sol mode, Sols 317 and 318 were planned together as a two-sol plan. For Sol 317, the science team elected to drive toward wheel tracks that Opportunity had made before entering Endurance Crater. The rover backed up about 5 meters (16.4 feet), performed some mid-drive imaging, and then continued another 10 meters (33 feet) to put the old rover tracks into the work volume of the robotic arm.

On this sol in 210 (2004 December 14):

MER-A Sol 338: Spirit drove 8 meters (26 feet) with 25 meters (82 feet) of commanded motion. Spirit saw up to 95-percent slip on some of the drive segments due to sandy terrain and the rover's tilt of 15 to 20 degrees.

—NASA/JPL

The Shorty Rogers song "Have You Hugged Your Martian Today," performed by the jazz octet Octobop, was released on their CD *After Dark*.

221 Mesha 28 **2025 August 24 12:00 UTC**

Ls 129.8° Curiosity Sol 4639
MSD 53908 Perseverance Sol 1614

On this sol in 174 (1937 April 01):

"Water for Mars" by Ross Rocklynne appeared in *Astounding Stories* (date approximate). The dashing Will Kair has a great plan to bring water to Mars. It's one the Empress of Mars uses all her powers to help implement, but can Will turn the wily plans of the exotic Bella Devlin to his own good? And what about Instar, the water-hater? Water on Mars will mean the end of his water-selling Empire. This story is replete with space battles, disintegration rays, force fields, and a polychromatic bridge.

—Ann Hardin, 2003

Cover art by Howard V. Brown. Interior art by H.W. Wesso.

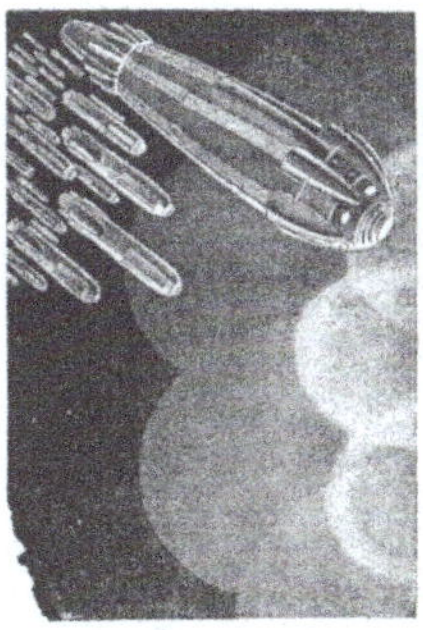

On this sol in 210 (2004 December 14):

MER-B Sol 318: This was another remote-sensing sol, during which *Opportunity* imaged its still-distant heat shield and conducted a miniature thermal emission spectrometer observation of the tracks.

On this sol in 210 (2004 December 15):

MER-A Sol 339: The rover team attempted another 25-meter (82-foot) drive. This was cut short at the start when the right rear wheel ingested a potato-sized rock. The rock apparently jammed between the inner part of the wheel and the drive mechanism, causing the drive current to exceed a pre-set limit, resulting in a safe motor stall.

—NASA/JPL

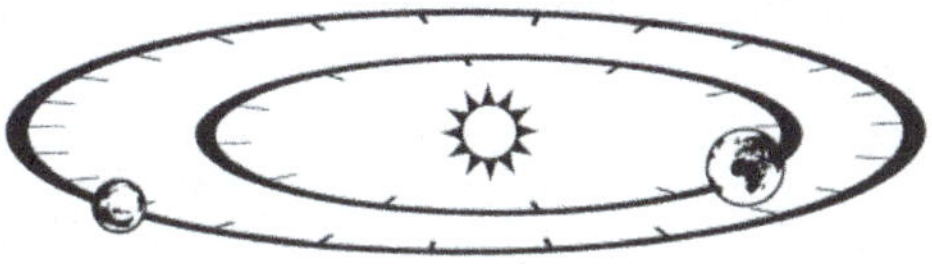

221 TAURUS

As seen from Mars, the Sun traverses the eastern sector of the constellation Aries, entering the western sector of the constellation Taurus later in the month. The identification of the latter constellation with a bull dates at least to the Chalcolithic Age (>7,000 years before present), and perhaps even to the Upper Paleolithic (>12,000 years before present).

The median one-way communications delay during this month is 19 minutes 3 seconds.

221 Taurus 01 **2025 August 25 12:40 UTC**

Ls 130.3° Curiosity Sol 4640
MSD 53909 Perseverance Sol 1605

Approximately the first sol of February on the Heinlein calendar.

On this sol in 181 (1950 June 02):

The motion picture *Rocketship X-M* was released. An astronaut crew on their way to the Moon are unexpectedly propelled by gravitational forces and end up on Mars instead.

—International Movie Database

On this sol in 184 (1956 January 23):

Journey into Space: The World in Peril, Episode 18 airdate. The asteroid fleet increases speed and overtakes Freighter #1. Jet records a warning to Earth, to be transmitted on a loop. Attempts to hypnotise the crew fail, so the Martian contacts the crew using their voices. Freighter #1 crashes onto the surface of a large asteroid.

—Wikipedia

On this sol in 210 (2004 December 15):

MER-B: Sol 319: *Opportunity* captured microscopic imager mosaics of both types of tracks, then drove about 40 meters (131 feet) closer to the heat shield, which will be examined carefully in future sols.

On this sol in 210 (2004 December 16):

MER-A Sol 340: *Spirit* made observations with the miniature thermal emission spectrometer to seek other rock targets similar to Wishstone. Turning the right rear wheel about 60 degrees successfully un-jammed the rock, but it remained inside the wheel.

Spirit successfully executed another small maneuver to get the right rear wheel out of hole and get the rock out of the wheel, but more steps will be required. The rover also used the panoramic camera and miniature thermal emission spectrometer to acquire information about nearby targets.

—NASA/JPL

221 Taurus 02 **2025 August 26 13:19 UTC**

Ls 130.8° Curiosity Sol 4641

MSD 53910 Perseverance Sol 1606

On this sol in 149 (1890 March 25):

Elmer Goodfellow "El" Brendel was born; American actor; starred in the motion picture *Just Imagine* (1930).

On this sol in 156 (1903 May 27):

Elwood Ullman was born; American screenwriter; wrote the screenplay for the motion picture *The Three Stooges in Orbit* (1962).

On this sol in 210 (2004 December 17):

MER-B Sol 320: Opportunity completed a variety of observations including imaging toward the heat shield. It then went into deep sleep overnight.

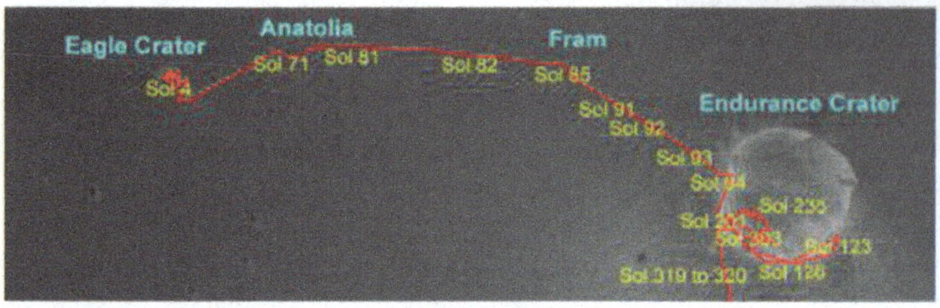

MER-A Sol 341: Sols 341, 342 and 343 (210 Taurus 02 through 04) were planned as a combined three-sol plan for *Spirit* that included observations with the miniature thermal emission spectrometer each sol. On Sol 341, *Spirit* used its microscopic imager and its Mössbauer spectrometer to examine disturbed soil in front of the rover. It switched to the alpha particle X-ray spectrometer overnight to gather more compositional information about the same target.

—NASA/JPL

221 Taurus 03 **2025 August 27 13:59 UTC**

Ls 131.3° Curiosity Sol 4642

MSD 53911 Perseverance Sol 1607

On this sol in 098 (1794 April 24):

William Herschel drew sketches of Mars. The opposition was also observed by Johann Schröter.

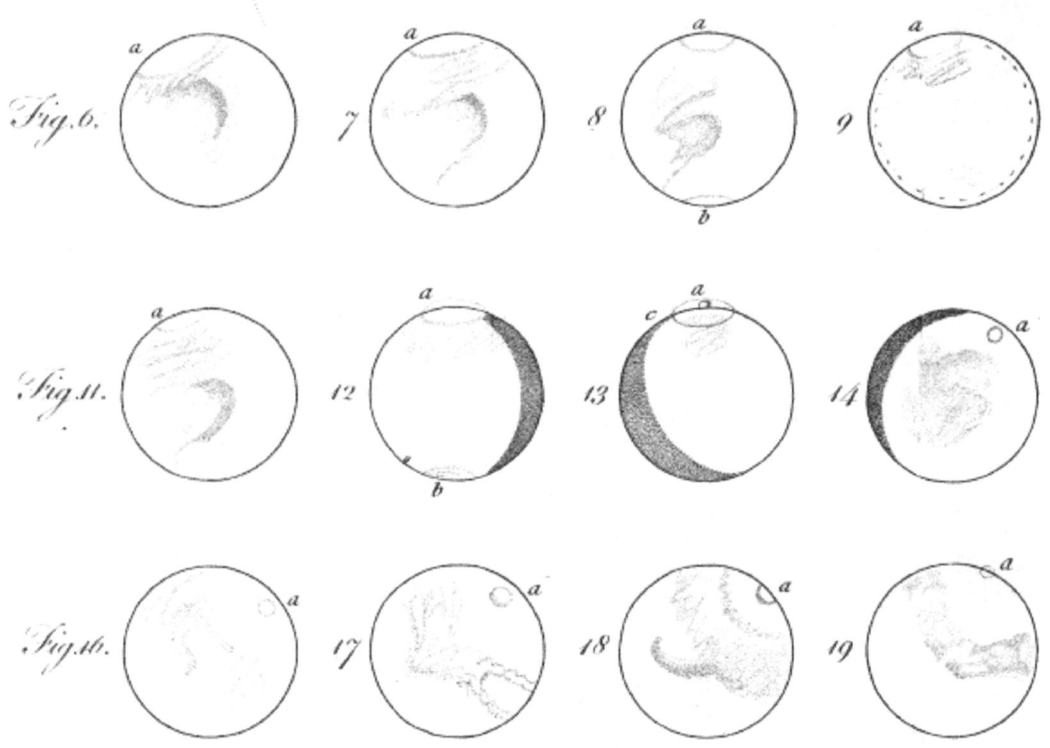

Herschel's drawings

On this sol in 210 (2004 December 18):

MER-B Sol 321: *Opportunity's* Sols 321, 322 and 323 were combined in a single activity plan for the Earth weekend. On the first sol of the three-sol plan, *Opportunity* looked at the front edge of its solar array with its microscopic imager to assess dust accumulation there. It then drove nearly 60 meters (about 197 feet) south toward the heat shield and acquired post-drive imagery to support planning of the next drive. *Opportunity* did not deep-sleep overnight so it could wake up to take advantage of a Mars Odyssey communications pass to return data during the night.

MER-A Sol 342: Sols 341, 342 and 343 (210 Taurus 02 through 04) were planned as a combined three-sol plan for *Spirit* that included observations with the miniature thermal emission spectrometer each sol. On Sol 342, *Spirit* performed a mid-day tool change back to the Mössbauer spectrometer.

—NASA/JPL

221 Taurus 04 — **2025 August 28 14:39 UTC**

Ls 131.8° — Curiosity Sol 4643
MSD 53912 — Perseverance Sol 1608

On this sol in 210 (2004 December 19):

MER-B Sol 322: *Opportunity* made some observations and then went into deep sleep overnight.

MER-A Sol 343: Sols 341, 342 and 343 (210 Taurus 02 through 04) were planned as a combined three-sol plan for *Spirit* that included observations with the miniature thermal emission spectrometer each sol. On Sol 343, the rover stowed the robotic arm and took images with the panoramic camera of targets that had been observed with the miniature thermal emission spectrometer. *Spirit* then performed a small maneuver but did not significantly change the position of the rock in the wheel.

—NASA/JPL

221 Taurus 05 — **2025 August 29 15:18 UTC**

Ls 132.3° — Curiosity Sol 4644
MSD 53913 — Perseverance Sol 1609

On this sol in 210 (2004 December 20):

MER-B Sol 323: *Opportunity* made more observations and did not deep-sleep overnight.

MER-A Sol 344: *Spirit* performed more remote sensing and did a maneuver that lifted the right rear wheel slightly out of a hole, but the rock remains partially in the wheel. The wheel is about one-third buried in the soft soil, making it difficult for the rock to escape until the wheel gets out of the hole.

—NASA/JPL

221 Taurus 06 — **2025 August 30 15:58 UTC**

Ls 132.8° — Curiosity Sol 4645
MSD 53914 — Perseverance Sol 1610

On this sol in 140 (1873 April 25):

Howard Roger Garis was born; American writer, co-author with Edward D. Stratemeyer (as Roy Rockwood) of *Through Space to Mars; or, The Longest Journey on Record* (1910).

On this sol in 149 (1890 March 29):

Karl R. Coolidge was born; American screenwriter and director; wrote the screenplay for the motion picture *Nan of the North: Missil from Mars* (1922).

On this sol in 210 (2004 December 21):

MER-B Sol 324: A healthy *Opportunity* had driven to within about 30 meters (about 98 feet) from the remains of the heat shield that hit the ground about 250 meters (820 feet) south of Endurance Crater.

Opportunity also used the panoramic camera and miniature thermal emission spectrometer to acquire information about nearby targets.

MER-A Sol 345: A potato-sized rock got caught in *Spirit's* right rear wheel on Sol 339 (210 Mesha 28), causing the wheel to stall and ending the drive for that sol. Small moves of the wheel on subsequent sols dislodged the rock, but the rock remains close to the wheel, so the team is planning small, careful steps to move the wheel away from the rock so it will not become jammed again. *Spirit* remains in excellent health.

Spirit finished work at a rock called Wishstone, then continued to make slow progress up Husband Hill. Wishstone is different than any rock *Spirit* previously studied either on the plains or in the hills. Scientists and engineers used the miniature thermal emission spectrometer to find similar rocks for further study.

Spirit successfully executed another small maneuver to get the right rear wheel out of hole and get the rock out of the wheel, but more steps will be required.

—NASA/JPL

221 Taurus 07 **2025 August 31 16:37 UTC**

Ls 133.3° Curiosity Sol 4646

MSD 53915 Perseverance Sol 1611

On this sol in 210 (2004 December 22):

MER-B Sol 325: *Opportunity* drove 27 meters (about 89 feet) backward, to West Point. It imaged the heat shield debris field from that vantage point. The engineers choose to occasionally drive the rover backward for convenience and to keep the wheel-motor lubrication more evenly distributed.

MER-A Sol 346: *Spirit* confirmed that it had dumped a potato-shaped rock that had been plaguing *Spirit's* right rear wheel. Confirmation came by comparison of before and after images from the rear hazard-avoidance camera. The total distance driven on Sol 346 was 0.33 meters (1.08 feet).

—NASA/JPL

221 Taurus 08 **2025 September 01 17:17 UTC**

Ls 133.8° Curiosity Sol 4647
MSD 53916 Perseverance Sol 1612

Mars arrival opportunity from Earth via a Type II trajectory.

Mars arrival opportunities from Earth via an EVM trajectory, and a Type IV trajectory.

Earth arrival opportunity from Mars via a Type II trajectory.

On this sol in 175 (1939 February 25):

"Nonstop to Mars" by Jack Williamson appeared in *Argosy* (date approximate). Here is the record of Lucky Leith's incredible flight; how he piloted his ship through the thunder between two worlds—to become the first Robinson Crusoe of space.

Interior art uncredited.

On this sol in 184 (1956 January 30):

Journey into Space: The World in Peril, Episode 19 airdate. Lemmy fails to contact Earth, but they are contacted by the Martian who tells them to enter the asteroid ship. The Martian reveals that he is the only one left of a race of giants, and explains his reasons for the invasion. A crew from the asteroid, including ex-freighter man Harding, forcibly remove the crew from Freighter #1.

—Wikipedia

On this sol in 208 (2001 March 21):

Mars Global Surveyor imaged the DiPietro-Molenaar Pyramid.

On this sol in 210 (2004 December 23):

MER-B Sol 326: This was the second sol of a two-sol plan. It was spent imaging the heat shield debris field.

MER-A Sol 347: *Spirit* observed selected targets with its panoramic camera and miniature thermal emission spectrometer.

—NASA/JPL

221 Taurus 09 **2025 September 02 17:57 UTC**

Ls 134.3° Curiosity Sol 4648
MSD 53917 Perseverance Sol 1613

The first sol of September, Martian year 3679 on the Levitt calendar.

On this sol in 205 (1995 July 31):

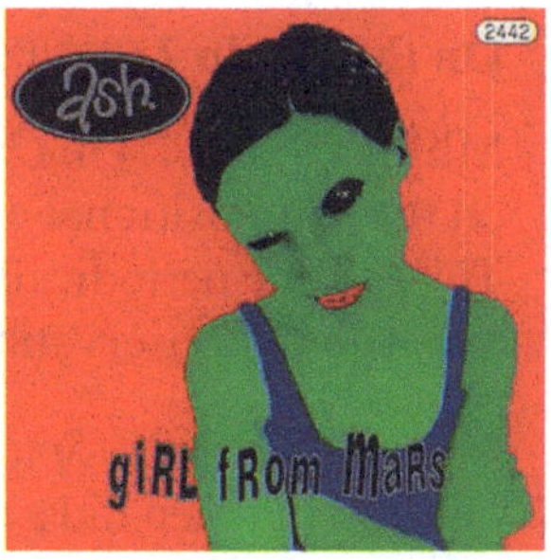

The song "Girl from Mars" by the Northern Irish punk rock band Ash was released on their vinyl LP and CD of the same name.

Do you remember the time I knew a girl from Mars?
I don't know if you knew that
Oh, we'd stay up late playing cards
Henri Winterman Cigars
And she never told me her name
I still love you, the girl from Mars

Sitting in our dreamy daze by the water's edge
On a cool summer night
Fireflies and the stars in the sky
Gentle glowing light from your cigarette

The breeze blowing softly on my face
Reminds me of something else
Something that in my memory has been misplaced
Suddenly all comes back
And as I look to the stars

I know that you are almost in love with me
I can see it in your eyes
Strange light shimmering over the sea tonight
And it almost blows my mind
And as I look to the stars

I remember the time I knew a girl from Mars
I don't know if you knew that
Oh, we'd stay up late playing cards
Henri Winterman Cigars
And she never told me her name
I still love you, the girl from Mars

I remember the time I knew a girl from Mars
I don't know if you knew that
Oh, we'd stay up late playing cards
Henri Winterman Cigars
And she never told me her name
I still love you, the girl from Mars

Surging through the darkness over the moonlight strand
Electricity in the air
Twisting all through the night on the terrace
Now that summer's here

Today I sleep in the chair by the window
It felt as if you'd returned
I thought that you were standing over me
When I woke there was no one there
I still love you, girl from Mars

Do you remember the time I knew a girl from Mars?
I don't know if you knew that
Oh, we'd stay up late playing cards
Henri Winterman Cigars
And she never told me her name

Do you remember the time I knew a girl from Mars?
I don't know if you knew that
Oh, we'd stay up late playing cards
Henri Winterman Cigars
And I still dream of you
I still love you, girl from Mars

On this sol in 210 (2004 December 24):

MER-B Sol 327: On *Opportunity's* Sols 327 to 329, the Mössbauer spectrometer was placed on the compositional calibration target for a series of observations over the Earth weekend. This is done periodically to calibrate the Mössbauer instrument. The rover continued routine atmospheric observations and remote sensing of the heat shield debris field.

MER-A Sol 348: *Spirit* took pictures of a target called Dreaming with the microscopic imager, and then did a tool change to the alpha particle X-ray spectrometer. *Spirit* also made some remote-sensing observations of targets with holiday theme names.

—NASA/JPL

221 Taurus 10 **2025 September 03 18:36 UTC**

Ls 134.8° Curiosity Sol 4649

MSD 53918 Perseverance Sol 1614

Midpoint between the northern solstice and the southward equinox, a cross-quarter sol analogous to Lammas Day (August 01) on Earth.

On this sol in 154 (1899 August 29):

Henrik Dahl Juve was born; American science fiction writer; author of *The Martian Revenge* (1930) and "In Martian Depths" (1932).

On this sol in 180 (1948 July 24):

Commander X-2 (later known as Marvin the Martian) debuted in animated short film "Haredevil Hare." Bugs Bunny is tricked into being the first rabbit shot into space. When he lands on the Moon he finds Commander X-2 and his dog K-9 set to blow up planet Earth.

—International Movie Database

On this sol in 210 (2004 December 25):

MER-B Sol 328: *Opportunity* continued a series of Mössbauer spectrometer observations on the compositional calibration target. The rover also continued routine atmospheric observations and remote sensing of the heat shield debris field.

221 Taurus 11 **2025 September 04 19:16 UTC**

Ls 135.3° Curiosity Sol 4650

MSD 53919 Perseverance Sol 1615

On this sol in 210 (2004 December 26):

MER-B Sol 329: *Opportunity* continued a series of Mössbauer spectrometer observations on the compositional calibration target. The rover also continued routine atmospheric observations and remote sensing of the heat shield debris field.

MER-A Sol 350: The plan was to drive *Spirit* 5 meters (16 feet) towards the rock named La Brea. However, the drive achieved less than a meter (3 feet) due to slippage.

—NASA/JPL

221 Taurus 12 **2025 September 05 19:55 UTC**

Ls 135.8° Curiosity Sol 4651

MSD 53920 Perseverance Sol 1616

As seen from Mars, the Sun enters the constellation Taurus.

On this sol in 210 (2004 December 27):

MER-B Sol 330: *Opportunity* used its panoramic camera to take images of the heat shield debris field, then drove 15 meters (about 49 feet) to a location called South Point for another look at the debris field.

This stunning image features the heat shield impact site of NASA's Mars Exploration Rover *Opportunity.* This is an approximately true-color mosaic of panoramic camera images taken through the camera's 750-, 530-, and 430-nanometer filters.

The mosaic was acquired on *Opportunity's* Sol 330, shortly after *Opportunity* arrived to investigate the site where its heat shield hit the ground south of Endurance Crater. On the left, the main heat shield piece is inverted and reveals its metallic insulation layer, glinting in the sunlight. The main piece stands about 1 meter tall (about 3.3 feet) and about 13 meters (about 43 feet) from the rover.

The other large, flat piece of debris near the center of the image is about 14 meters (about 46 feet) away. The circular feature on the right side of the image is the crater made by the heat shield's impact. It is about 2.8 meters (9.2 feet) in diameter but only about 5 to 10 centimeters (about 2 to 4 inches) deep. The crater is about 6 meters (about 20 feet) from *Opportunity* in this view. Smaller fragments and debris can be seen all around the impact site.

The impact excavated a large amount of reddish subsurface material. Darker materials cover part of the crater's flat floor and have formed a streak or jet of material pointing toward the two largest heat shield fragments.

MER-A Sol 351: *Spirit* attempted to drive again. This drive involved a series of rearward and forward arcs to get to more favorable terrain. Most of the slip occurred during the forward arcs. The estimated slip on the rearward arcs was 15 per cent, but the estimated slip on the forward arcs was 39 per cent. The destination, a rock called Dick Clark, was still about 4.2 meters away (14 feet).

—NASA/JPL

221 Taurus 13 **2025 September 06 20:35 UTC**
Ls 136.3° Curiosity Sol 4652
MSD 53921 Perseverance Sol 1617

On this sol in 141 (1875 March 20):

Roger Burton Morrison was born; American pulp artist; created the interior art for the serialization of *The Chessmen of Mars* by Edgar Rice Burroughs (1922).

On this sol in 210 (2004 December 28):

MER-B Sol 331: *Opportunity* drove roughly 10 meters (33 feet) to approach the flank portion of the heat shield remains. The heat shield broke into two main pieces when it hit the ground. The flank is the smaller of those portions.

MER-A Sol 352: The incomplete drive on the previous sol had left the rover team with a rock of interest right between *Spirit's* front wheels. *Spirit* examined a target called Bubbles with its microscopic imager, then changed tools to the alpha particle X-ray spectrometer. *Spirit* also made sky observations with its thermal emission spectrometer and panoramic camera.

—NASA/JPL

221 Taurus 14 **2025 September 07 21:14 UTC**
Ls 136.8° Curiosity Sol 4653
MSD 53922 Perseverance Sol 1618

On this sol in 184 (1956 February 06):

Journey into Space: The World in Peril, Episode 20 airdate. The crew are shown to luxurious quarters aboard the asteroid, and cruise to Earth as part of the invasion fleet. They are ordered to the Martian's quarters, where they are dismayed to discover that Earth has not followed their instruction to shut down all television channels. Depressed by their failure, they watch as the Martian fleet begins to orbit the Earth. As they wait for the Martian's broadcast, the set goes dead. They are overjoyed to learn that the channels have at last been shut down. The Martian peacefully accepts defeat and allows all those in his command to return to Earth. A considerable number, however, choose to remain in the Martian's fleet, and to establish a perfect civilisation in another solar system, free from persecution, hunger, misgovernment and war.

—Wikipedia

On this sol in 210 (2004 December 29):

MER-B Sol 332: *Opportunity* is healthy and has reached the site where its heat shield hit the ground. *Opportunity* will make detailed observations of the heat shield's remains, weather permitting. *Opportunity* experienced its first dust storm since landing, which has affected the amount of energy *Opportunity* gets each sol. When *Opportunity* landed nearly one Earth year ago, a dust storm was subsiding and the atmosphere had an opacity of 0.9

(the higher the number, the murkier the skies). Since then, the opacity had improved significantly and was roughly 0.5 on Sol 327 (210 Taurus 09). On Sol 328 (210 Taurus 10) the opacity jumped to 0.6 then to 0.8, 1.2, and 1.25 on Sols 329-331 (210 Taurus 11-13). As of Sol 332 it is at 1.2 and dropping. Images from *Mars Global Surveyor* orbiter have confirmed the presence of a few small dust storms in the region. The energy intake has decreased roughly 30 percent, leaving *Opportunity* with less energy for operations and communications but still enough, with comfortable margin, to continue with the plan to investigate the heat shield remains. The dust storms will be monitored carefully using *Opportunity's* own instruments and images from *Mars Global Surveyor*. The team will also be walking through low-energy contingencies should they become necessary.

Opportunity made its final approach to the flank portion of the heat shield wreckage in preparation for close-up inspection of the heat shield material over the New Year's holiday weekend. The drive brought *Opportunity's* odometer total to 2,051 meters (1.27 miles).

MER-A Sol 353: *Spirit* continued inspecting the rock Champagne and performed a Mössbauer spectrometer integration on two Champagne targets.

—NASA/JPL

221 Taurus 15 **2025 September 08 21:54 UTC**
Ls 137.3° Curiosity Sol 4654
MSD 53923 Perseverance Sol 1619

The first sol of the fourth quarter of Summer on the Aitken calendar.

On this sol in 146 (1884 August 16):

Hugo Gernsbacher (Hugo Gernsback) was born; Luxembourgish-American publisher and editor of *Amazing Stories, The Electrical Experimenter,* and *Wonder Stories;* author of "How the Martian Canals Are Built" (1916), "Martian Amusements" (1916), "Martian Atmosphere Plants" (1917); namesake of the Hugo Award for science fiction.

On this sol in 210 (2004 December 30):

MER-B Sol 333: *Opportunity* spent the sol monitoring the opacity of the atmosphere and performing almost three hours of remote sensing. Atmospheric opacity peaked on Sol 321 (210 Taurus 03) at 1.25. On Sol 333 it was down to 0.97 and decreasing, indicating that the sky was clearing. The amount of power generated by *Opportunity's* solar panels increased from 546 watt-hours on Sol 321 to 630 watt-hours on Sol 333.

MER-A Sol 354: *Spirit* used its rock abrasion tool to brush the dust off a target on Champagne, then used the microscopic imager and the alpha particle X-ray spectrometer to analyze the dust-free spot.

—NASA/JPL

221 Taurus 16 **2025 September 09 22:34 UTC**

Ls 137.8° Curiosity Sol 4655

MSD 53924 Perseverance Sol 1620

On this sol in 152 (1895 December 01):

Mars by Percival Lowell was published (date approximate). Together with *Mars and Its Canals* (1906), and *Mars as the Abode of Life* (1908), Lowell more than anyone else popularized the long-held belief that these markings showed that Mars sustained intelligent life forms. His works include a detailed description of what he termed the "non-natural features" of the planet's surface, including especially a full account of the "canals," single and double; the "oases," as he termed the dark spots at their intersections; and the varying visibility of both, depending partly on the Martian seasons. He theorized that an advanced but desperate culture had built the canals to tap Mars's polar ice caps, the last source of water on an inexorably drying planet. While this idea excited the public, the astronomical community was skeptical. Many astronomers could not see these markings, and few believed that they were as extensive as Lowell claimed.

—Wikipedia

Nevertheless, Lowellian Mars, a world where humans could live and breathe, became the setting for hundreds of science fiction tales, including the works of Edgar Rice Burroughs, Leigh Brackett, Robert A. Heinlein, Ray Bradbury, Michael Moorcock, and Lin Carter. A century after Lowell's work, fiction set on a terraformed Mars can be seen as a subgenre that makes Mars plausibly Lowellian for a present-day readership.

On this sol in 173 (1935 June 01):

"Mars Colonizes" by Miles J. Breuer appeared in *Marvel Tales* (date approximate). Rather than attempt to conquer Earth, Martians might do better to just move to Earth and buy up real estate.

Part 3 of *The Titan* by P. Schuyler Miller also appeared in *Marvel Tales* (date approximate).

On this sol in 210 (2004 December 31):

MER-B Sol 324: *Opportunity* drove nearly 90 meters (about 295 feet), bringing it within 30 meters (98 feet) of the heat shield. This was a record length for a drive that did not use the rover's auto-navigation software. It was enabled by the team's ability to identify a hazard-free route in images that the navigation camera and panoramic camera had taken of the flat terrain to be covered in the drive. The rover also made use of low-level hazard-detection software that reads rocker-bogie ("rover leg") position and rover tilt to halt a drive if the rover encounters a hazard. This drive brings the odometer 1,998 meters (1.24 miles).

MER-A Sol 355: *Spirit* used the rock abrasion tool to grind into Champagne, then placed the alpha particle X-ray spectrometer in the hole for an overnight observation.

—NASA/JPL

221 Taurus 17 **2025 September 10 23:13 UTC**

Ls 138.3° Curiosity Sol 4656
MSD 53925 Perseverance Sol 1621

On this sol in 149 (1890 April 09):

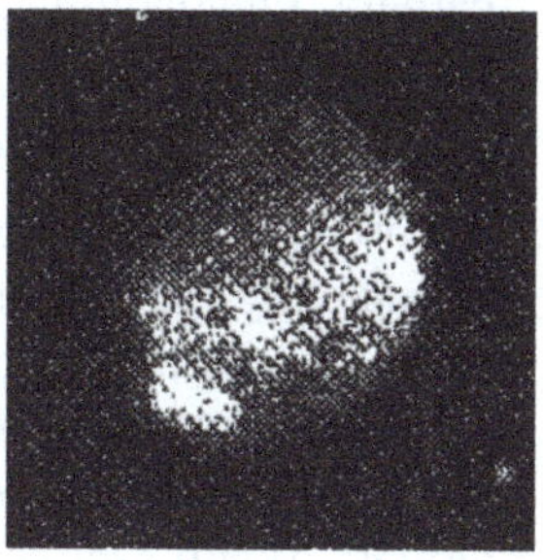

Harvard University astronomers photographed Mars.

According to William H. Pickering, the Martian date was August 08.

On this sol in 155 (1901 July 24):

Igor’ Vladimirovich Ilyinskij (И́горь Влади́мирович Ильи́нский) was born; Russian/Soviet actor; starred in the motion picture *Аэлита - королева Марса (Aelita: Queen of Mars)* (1924).

On this sol in 210 (2005 January 01):

MER-B Sol 335: *Opportunity* started taking data with the alpha particle X-ray spectrometer early in the morning, napped several hours, awoke and turned off that spectrometer, performed another hour of remote sensing, and then went to sleep for the night.

MER-A Sol 356: *Spirit* continued using its alpha particle X-ray spectrometer on the hole drilled by the rock abrasion tool hole.

—NASA/JPL

221 Taurus 18 **2025 September 11 23:53 UTC**

Ls 138.8° Curiosity Sol 4657
MSD 53926 Perseverance Sol 1622

On this sol in 149 (1890 April 10):

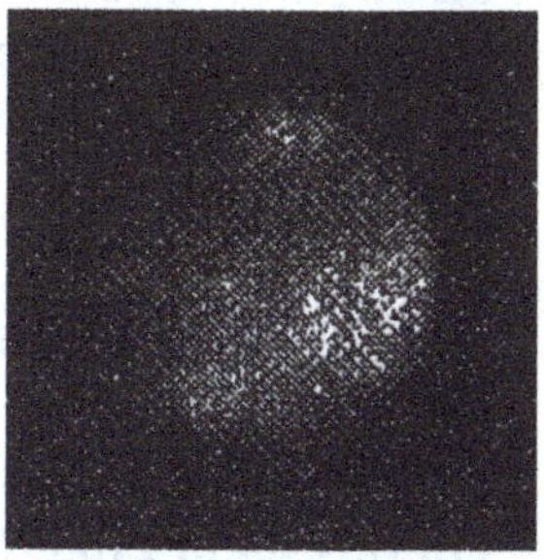

Harvard University astronomers photographed Mars.

According to William H. Pickering, the Martian date was August 09.

On this sol in 210 (2005 January 02):

MER-B Sol 336: *Opportunity* collected another two hours of alpha particle X-ray spectrometer data on the capture magnet and made remote-sensing observations for an hour.

On this sol in 210 (2005 January 03):

MER-A Sol 357: *Spirit* changed tools to the Mössbauer spectrometer and placed it in the rock abrasion tool hole for a long integration.

—NASA/JPL

221 Taurus 19 **2025 September 13 00:32 UTC**

Ls 139.3° Curiosity Sol 4658
MSD 53927 Perseverance Sol 1623

On this sol in 210 (2005 January 03):

MER-B Sol 337: *Opportunity* examined the flank portion of the heat shield wreckage with the microscopic imager, imaged the capture magnet, then stowed the arm and began its drive for the day. *Opportunity* backed up 2 meters (about 6.6 feet), turned and drove 10 meters (33 feet) to East Point. At East Point, Opportunity performed 30 minutes of imaging, capturing a stereo image of the heat shield and a 360-degree view with the navigation camera. *Opportunity* then drove another 12.5 meters (41 feet) to a standoff point facing the heat shield seal.

The team experienced a problem with onboard file deletion on Sol 337. There was a bad parameter in one command, so *Opportunity* rejected the entire set of commands. The result of rejecting the command file was that *Opportunity's* flash memory filled up and the flight software began deleting lowest-priority data products. Approximately 150 megabits of stored data products were lost. Because many images were being acquired and processed at the same time auto-deleting was taking place, *Opportunity's* computer was running very slowly. When *Opportunity* attempted to perform a "Get Fine Attitude," (a command that updates *Opportunity's* knowledge of its tilt and orientation), it timed out. *Opportunity* monitors itself and cancels things that take longer than planned, the same way home computers cancel attempts to access the internet if the server doesn't respond within a certain time. Due to the time-out, *Opportunity* responded that it didn't know its exact attitude (setting the surface attitude pointing and positioning — SAPP — knowledge to "unknown"). As a result of this, subsequent commands for using the miniature thermal mission spectrometer images were rejected. Pointing the spectrometer at the Sun would severely damage it, so in order for *Opportunity* to point the instrument, it must know its attitude.

On this sol in 210 (2005 January 04):

MER-A Sol 358: *Spirit* took microscopic images of the rock abrasion tool hole, then stowed the rover arm. Engineers then attempted a short drive that would have allowed *Spirit* to image the rock abrasion tool hole from a distance. The drive did not reach the intended position because of slippage.

—NASA/JPL

221 Taurus 20 **2025 September 14 01:12 UTC**

Ls 139.9° Curiosity Sol 4659

MSD 53928 Perseverance Sol 1624

On this sol in 210 (2005 January 04):

MER-B Sol 338: This became a restricted-operations sol for *Opportunity* due to a long latency in receiving relayed data. Telemetry was delayed more than seven hours, so *Opportunity* spent the sol performing more than four hours of remote sensing. The operations team was able to react to the failed Get Fine Attitude procedure on Sol 337 by creating a real-time command to reset the surface attitude pointing and positioning knowledge. The command was sent as part of the Sol 338 uplink and worked as planned. *Opportunity* also performed both a left-eye and right-eye Get Fine Attitude.

On this sol in 210 (2005 January 05):

MER-A Sol 359: Engineers continue to have difficulty getting the rover to make significant progress toward a ridgeline destination due to high slippage that *Spirit* experiences on the sandy, sloped terrain. *Spirit* has been in roughly the same spot for the past 30 sols. Since temporarily getting a potato-sized rock caught in the right rear wheel on Sol 339 (210 Mesha 28), engineers have been careful to monitor the slip as *Spirit* drives. If the wheels slip too much, engineers stop the drive to avoid the possibility of picking up another rock as *Spirit* spins and digs with its wheels.

Sol 359 was a restricted sol, meaning that engineers had to plan this sol without knowing the results of the Sol 358 (210 Taurus 19) sequence. Consequently, it was a relatively simple day that included routine atmospheric science and miniature thermal emission spectrometer observations.

—NASA/JPL

221 Taurus 21 **2025 September 15 01:52 UTC**
Ls 140.4° Curiosity Sol 4660
MSD 53929 Perseverance Sol 1625

On this sol in 171 (1931 September 01):

"The Disc-Men of Jupiter" by Manly Wade Wellman appeared in *Wonder Stories* (date approximate). Mars is dying and will not be habitable much longer. As a result, Martian culture has turned to wild, fin de siècle decadence, with outlandish drugs, sensation-enhancing rays, and general hedonism.

—Bleiler and Bleiler, 1998

Interior art by Marco Marchioni.

"The Planet Entity" (later titled "Seedling of Mars") by Clark Ashton Smith (based on a plot by E.M. Johnston) appeared in *Wonder Stories Quarterly* (date approximate). It was in the fall of 1947, three days prior to the annual football game between Stanford and the University of California, that the strange visitor from outer space landed in the middle of the huge stadium at Berkeley where the game was to be held.

Interior art by Marco Marchioni.

On this sol in 210 (2005 January 05):

MER-B Sol 339: NASA's Mars Exploration Rover *Opportunity* has found an iron meteorite on Mars (now known as Heat Shield Rock), the first meteorite of any type ever identified on another planet. The pitted, basketball-size object is mostly made of iron and nickel. Readings from spectrometers on the rover determined that composition. *Opportunity* used its panoramic camera to take the images used in this approximately true-color composite on the rover's 339th sol. This composite combines images taken through the panoramic camera's 600-nanometer (red), 530-nanometer (green), and 480-nanometer (blue) filters.

Sol 339 was another "image, drive, image, drive" sol. *Opportunity* performed 30 minutes of pre-drive imaging, drove about 10 meters (33 feet), performed 80 minutes of mid-drive imaging, drove another 13 meters (about 43 feet) toward the charred side of the heat shield, and performed 30 minutes of post-drive imaging. All operations worked as planned, leaving *Opportunity* in position to approach the heat shield to perform microscopic imaging over the weekend.

On this sol in 210 (2005 January 06):

MER-A Sol 360: This sol was a repeat of Sol 358's drive. *Spirit* performed an "S" turn and then drove straight for about 4 additional meters (13 feet). This was a refreshingly good result for the tough terrain where *Spirit* was located. *Spirit* experienced as much as 48 percent slippage during the drive, but ended up moving a total distance of 9.7 meters (31.8 feet). *Spirit* also performed 20 minutes of post-drive imaging.

The opacity of the sky — or how much light does not shine through the atmosphere — has been higher than normal in the past week due to local dust disturbances. Higher opacity means less energy for the solar-powered rover. However, despite the recent increase in opacity, Spirit has had adequate energy (with a safe margin) to continue normal operations.

The rover's power bus return is a collection of wires designed to carry current back to the rover power source (battery or solar array). Electrical current between *Spirit's* chassis and the power bus return changed on Sol 342 (210 Taurus 03) from zero volts to 0.1 volts. The small change in voltage coincided with powering of heater circuits on the instrument deployment device (or rover arm). This confirms that there is a short (unexpected metal-to-metal contact) between the power bus return and the chassis, as had been suspected since 210 Aries. The short is somewhere on the return side of the rover arm heater circuits. The presence of this short does not affect daily operations. It does, however, take away one layer of protection should *Spirit* have a short to the chassis somewhere else on *Spirit.*

With one eye on the weather, *Spirit* continued work on Husband Hill, making detailed observations of a rock called Champagne, using the full suite of instruments.

—NASA/JPL

221 Taurus 22 **2025 September 16 02:31 UTC**

Ls 140.9° Curiosity Sol 4661
MSD 53930 Perseverance Sol 1626

On this sol in 149 (1890 April 15):

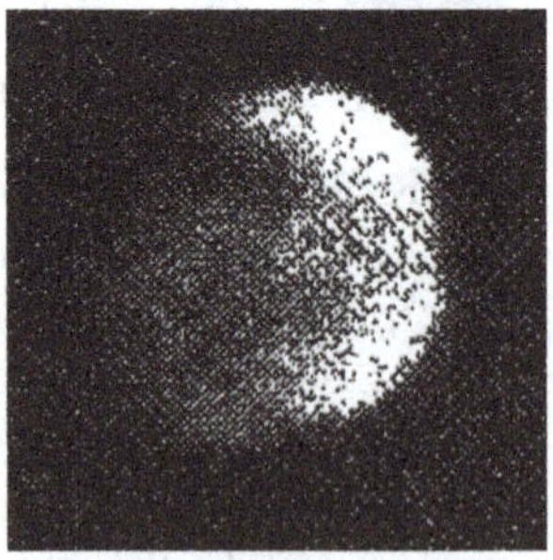

Harvard University astronomers photographed Mars.

According to William H. Pickering, the Martian date was August 14.

On this sol in 210 (2005 January 06):

MER-B Sol 340: A dust storm that affected *Opportunity* the previous week has slowly receded, allowing increasing solar exposure. To conserve energy, *Opportunity* has been going into the deep-sleep mode every night, but as power continues to improve, the team is planning to resume using some early-morning *Mars Odyssey* communication passes to reduce a backlog of unsent telemetry.

Opportunity continues to be in excellent health as the *Opportunity* team looks forward to the 2005 January 24 Earth anniversary of *Opportunity's* landing.

On this sol in 210 (2005 January 07):

MER-A Sol 361: This was a restricted sol for *Spirit* due to a late downlink of data needed for more extensive planning. That ruled out any driving or use of the instrument deployment device. *Spirit* completed almost 2 hours of remote sensing observations using the panoramic camera and miniature thermal emission spectrometer.

—NASA/JPL

221 Taurus 23 **2025 September 17 03:11 UTC**

Ls 141.4° Curiosity Sol 4662
MSD 53931 Perseverance Sol 1627

Earth departure opportunity to Mars via an EEM trajectory.

Mars arrival opportunity from Earth via a Type II trajectory.

On this sol in 143 (1879 January 02):

Avigdor "Victor" Rousseau Emanuel was born; British science fiction writer; author of "The Lord of Space" (1930).

On this sol in 210 (2005 January 07):

MER-B Sol 341: *Opportunity* Sols 341 through 343 were combined in a three-sol plan for the Earth weekend. On Sol 341, *Opportunity* used a morning *Mars Odyssey* pass for a communications relay at about 4 a.m. local solar time and then slept until solar-array wakeup at about 8:45 a.m. After another short nap, *Opportunity* did a bit of remote sensing and received its new commands for the sol. *Opportunity* deployed its robotic arm and acquired 96 microscopic images of the fractured edge of the heat shield. In the afternoon it used its Mössbauer spectrometer to analyze dust on the science filter magnet. It used the deep-sleep mode overnight.

On this sol in 210 (2005 January 08):

MER-A Sol 362: Sol 362 and 363 were planned in a single planning cycle. Sol 362 was another remote sensing sol.

—NASA/JPL

221 Taurus 24 **2025 September 18 03:50 UTC**

Ls 141.9° Curiosity Sol 4663
MSD 53932 Perseverance Sol 1628

On this sol in 162 (1914 October 01):

Donald Allen Wollheim (David Grinnell) was born; American science fiction writer and editor; co-author of *The Martians Are Coming* (1941), author of *Secret of the Martian Moons* (1955) and *The Martian Missile* (1959).

On this sol in 210 (2005 January 08):

MER-B Sol 342: *Opportunity* woke from deep sleep at about 7 a.m. local solar time and restarted the Mössbauer integration on the magnet. It was a light day of activity with afternoon remote sensing, an evening *Mars Odyssey* relay pass, and then deep-sleeping overnight.

On this sol in 210 (2005 January 09):

On Sol 363, *Spirit* attempted a 10-meter (33-foot) drive. *Spirit* accomplished about 3 meters (10 feet) of the drive before stopping due to high wheel slippage. *Spirit* experienced slippage of up to 98 percent on portions of this drive. *Spirit* then performed 20 minutes of imaging.

—NASA/JPL

221 Taurus 25 **2025 September 19 04:30 UTC**

Ls 142.4° Curiosity Sol 4664

MSD 53933 Perseverance Sol 1629

On this sol in 210 (2005 January 09):

MER-B Sol 343: *Opportunity* restarted Mössbauer integration on the magnet and completed afternoon remote sensing before deep-sleeping.

On this sol in 210 (2005 January 10):

MER-A Sol 364: The rover planners devised a strategy that would attempt to recover when *Spirit* experienced slippage of greater than 60 percent. The drive sequence was complex and *Spirit's* planning team had a tight planning cycle, so engineers came painfully close to missing the uplink window. The payoff for the long day was a very good drive. *Spirit* accomplished 7 meters (23 feet).

—NASA/JPL

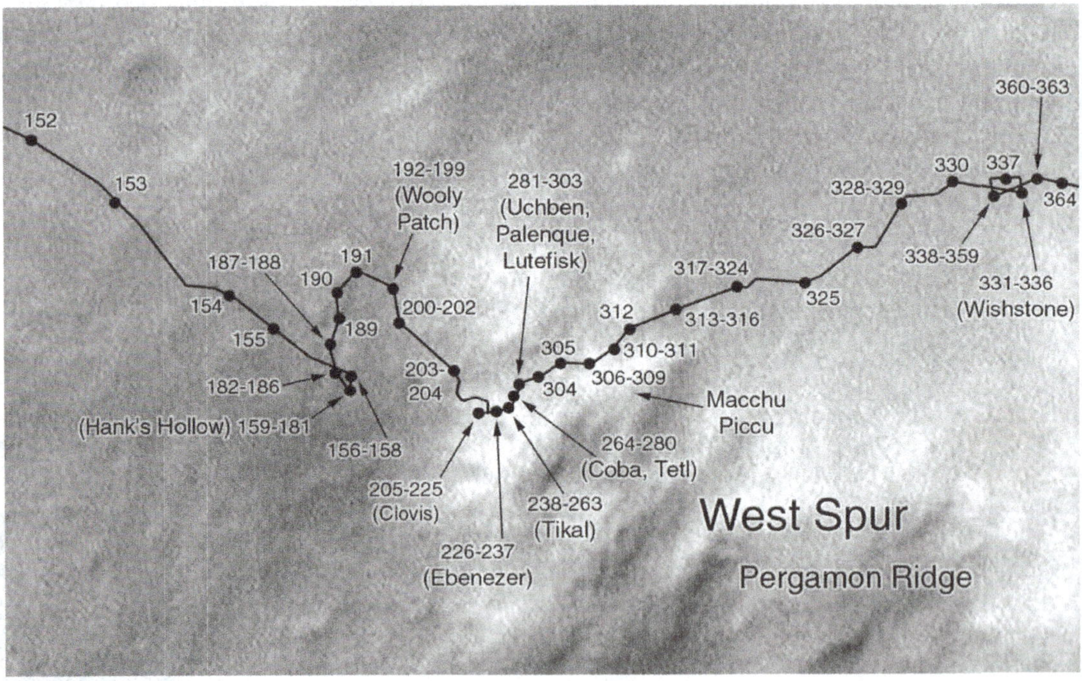

221 Taurus 26 **2025 September 20 05:09 UTC**

Ls 142.9° Curiosity Sol 4667
MSD 53934 Perseverance Sol 1630

On this sol in 144 (1880 November 23):

Basil Dickey was born; American writer; co-wrote the screenplay for the motion picture *The Purple Monster Strikes* (1945).

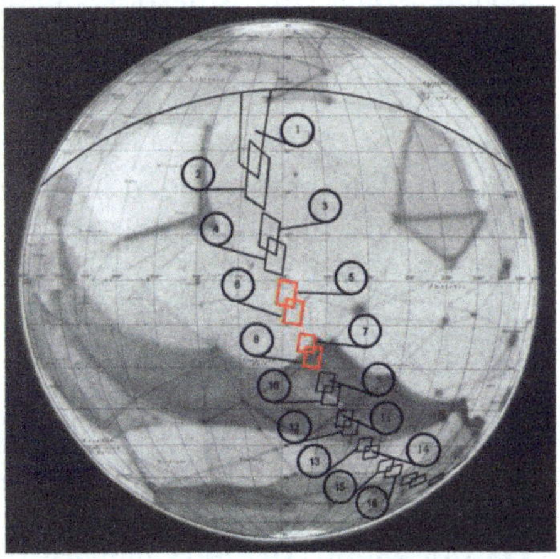

On this sol in 189 (1965 July 15):

Mariner 4 achieved the first flyby of Mars, inaugurating the Telemetric Era of Martian history, during which Mars was investigated via robotic spacecraft, and the Reconnaissance Period of sporadic telemetric investigation of Mars. *Mariner 4* passed within 9,846 kilometers (6,118 miles) and transmitted 22 images of a cratered surface.

Frames 1D through 4B are presented in *The Martian Almanac 220,* Volume 1.

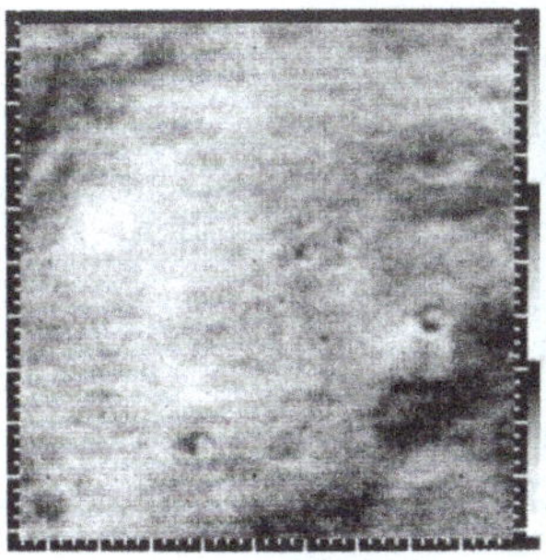

Mariner 4 frame 05B. This image taken from 14,300 km shows some craters in southern Amazonis Planitia. The image is 280 km across, 375 km from top to bottom and north is at 11:00. The sun is 19 degrees from vertical so some detail can be seen, but the shading is still rather subdued. The image is centered at 2° S, 181° E.

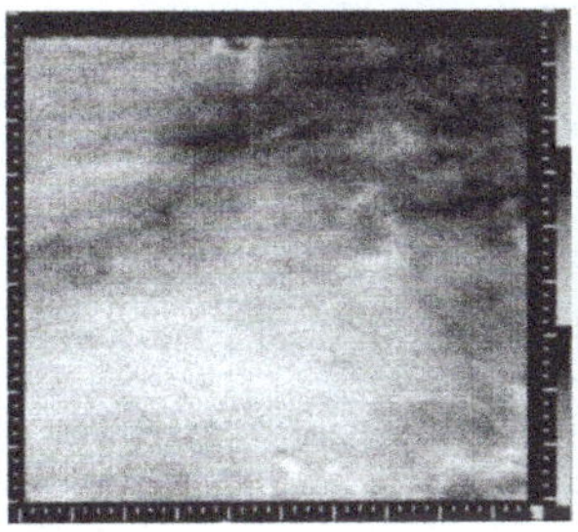

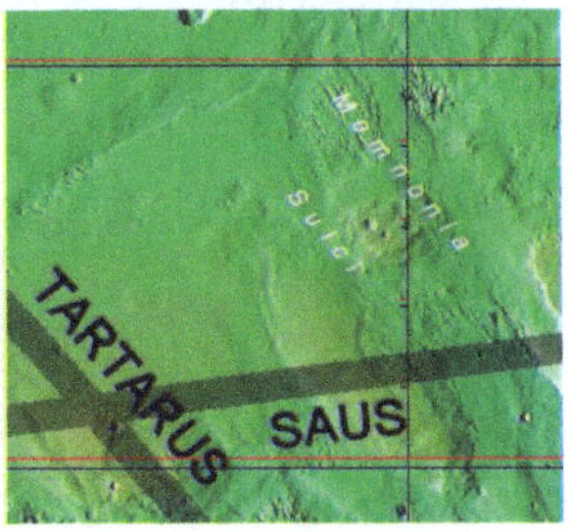

Mariner 4 frame 06E. Image showing a dark feature in the southern Amazonis Planitia. The part of this feature at upper left is also visible in the lower right of Image 5. The image was taken from 14,100 km and the frame is 273 km across and 349 km vertically. North is at 11:00. Central Latitude/Longitude (deg): 05.8° S, 182.9° E.

Mariner 4 frame 07B. With a range of 13,600 km and a sun angle of 29 degrees from zenith, this *Mariner 4* image was the first picture showing unambiguous craters on the surface of Mars. The area, 262 x 310 km, is a heavily cratered region south of Amazonis Planitia, centered at 14° S, 186° E. North is at about 11:00.

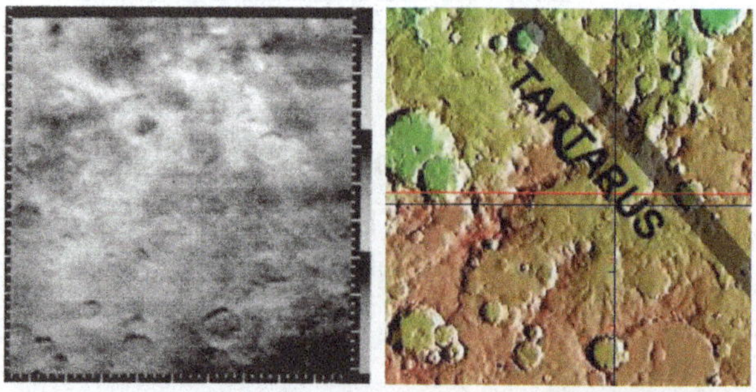

Mariner 4 frame 08E. This image confirmed the craters found in Image 7. (Image 8 overlaps in its top left corner with the lower right of image 7.) The image was taken from 13,400 and covers 255 km by 296 km. The two craters at the center of the image are about 32 km in diameter. North is about 11:00. Central Latitude/Longitude (deg): -17.1°, 187.4° E.

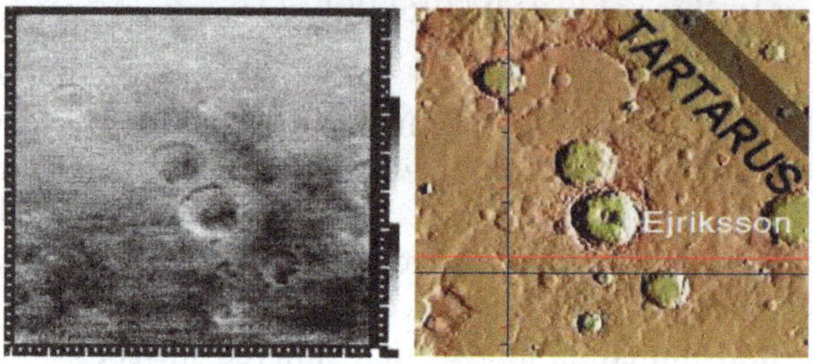

On this sol in 210 (2005 January 10):

MER-B Sol 344: *Opportunity* stowed its instrument deployment device (robotic arm) and backed up about 4 meters (13 feet) before acquiring a panorama of the heat shield and other remote sensing. The rover did not deep-sleep overnight in order utilize the morning *Mars Odyssey* relay and return as much data as possible. The flash memory is relatively full.

On this sol in 210 (2005 January 11):

MER-A Sol 365: *Spirit* performed another uphill drive. The rover drove 9 meters (29 feet) and performed 20 minutes of post-drive imaging.

—NASA/JPL

221 Taurus 27 **2025 September 21 05:49 UTC**
Ls 143.4° Curiosity Sol 4666
MSD 53935 Perseverance Sol 1631

On this sol in 176 (1941 February 01):

"The Winking Lights of Mars" by Otto Binder (as Gordon A. Giles) appeared in *Amazing Stories*. The linking of two worlds hinged on the result of the astronomers' observations. Would the Winking Lights be seen?

Interior art by Jay Jackson.

"The Teacher from Mars" by Earl and Otto Binder (as Eando Binder) appeared in *Thrilling Wonder Stories* (date approximate). The old professor from the crimson planet feared Earth's savagery—until humanity taught him a profound secret!

Interior art uncredited.

On this sol in 210 (2005 January 11):

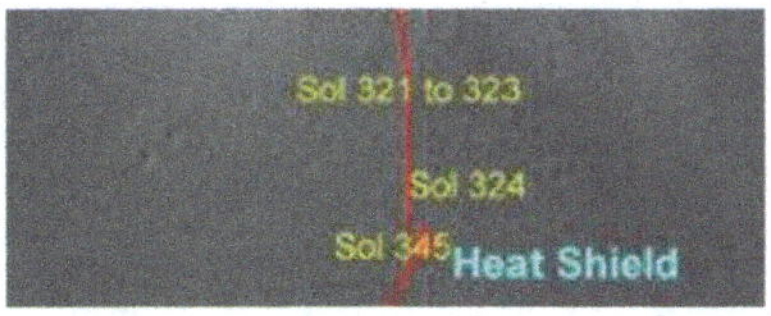

MER-B Sol 345: *Opportunity* acquired additional navigation-camera images of the heat shield to support a future re-approach for additional microscopic imaging. It then turned and drove north toward Heat Shield Rock, which has pits in its surface. The rover traversed about 10 meters (33feet) and arrived at the desired 1-meter (about 3.3 feet) standoff distance to acquire remote sensing of the rock. *Opportunity* used the energy-conserving deep-sleep mode overnight.

On this sol in 210 (2005 January 12):

MER-A Sol 366: Currently *Spirit* is approximately 50 meters (164 feet) from a target called Larry's Lookout on a ridge line in the Columbia Hills. During the period from Sol 360 (210 Taurus 21) through Sol 366, engineers focused on maximizing the amount of time *Spirit* could drive every sol with limited power. The driving is slow and difficult; *Spirit* is encountering many rock obstacles and patches of soft sand that are causing *Spirit* to either slip or dig in. Just when it looked like *Spirit* might not be able to reach Larry's Lookout, the rover had three successful drive sols.

The plan on Sol 366 was to continue the uphill drive toward Larry's Lookout, employing all the latest rover tricks. The drive was successful as *Spirit* traversed more than 12 meters (39 feet) with slip averaging less than 15 degrees. *Spirit* performed 30 minutes of post-drive observations.

—NASA/JPL

221 Taurus 28 **2025 September 22 06:29 UTC**

Ls 143.9° Curiosity Sol 4667

MSD 53936 Perseverance Sol 1632

On this sol in 210 (2005 January 12):

MER-B Sol 346: *Opportunity* is healthy. It acquired microscopic images of the fractured edge of the heat-shield wreckage and began a detailed investigation of an intriguing, pitted rock a few meters to the north, called Heat Shield Rock. The team continues to closely monitor orbital images for dust storms. The opacity of the atmosphere above *Opportunity* has averaged 0.75 with a slight downward (clearing) trend over the past week.

On this sol in 210 (2005 January 13):

MER-A Sol 367: *Spirit* traversed about 14 meters (46 feet) up the steep hillside toward the ridge and a target named Larry's Lookout. The average slippage during the drive is estimated at 14 percent, indicating much firmer footing than previous drives.

—NASA/JPL

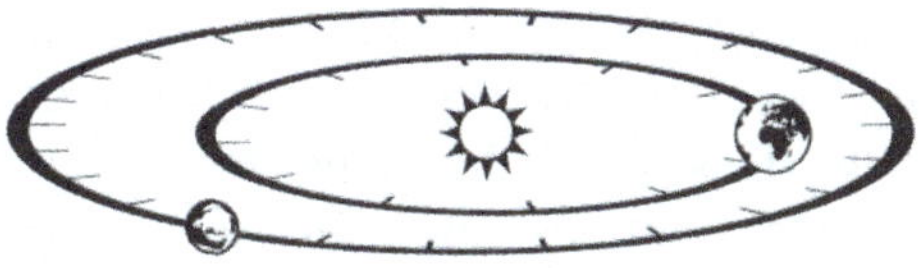

221 RISHABHA

As seen from Mars, the Sun traverses the central sector of the constellation Taurus. Rishabha (ऋषभ) is one of the avatars of Vishnu (विष्णु) in the Bhagavata Purana (भागवत पुराण). Rishabha is also found in Vedic literature, where it means "bull" and is an epithet for Shiva (शिव). A variant of the name, Vrisha (वृषा), is a month in the Hindu solar calendar that originally occurred when the Sun passed through the constellation Taurus as seen from Earth.

The median one-way communications delay during this month is 19 minutes 43 seconds.

221 Rishabha 01 **2025 September 23 07:08 UTC**

Ls 144.5° Curiosity Sol 4668

MSD 53937 Perseverance Sol 1633

Approximately the first sol of the unnamed month between February and March on the Heinlein calendar.

On this sol in 145 (1882 October 13):

Isaac R. Nathanson was born; American science fiction writer; author of *Moon People of Jupiter* (1931).

On this sol in 157 (1905 May 10):

Antonio Alejandro Schomburg y Rosa (Alex Schomburg) was born; American commercial artist, comic book artist, and painter; created the cover art for *Secret of the Martian Moons* by Donald A. Wollheim (1955).

On this sol in 156 (1903 June 23):

Anthony Veiller was born; American film producer, screenwriter, and director; co-wrote the screenplay for the motion picture *Red Planet Mars* (1952).

On this sol in 210 (2005 January 13):

MER-B Sol 347: *Opportunity* unstowed its instrument deployment device (robotic arm) to take microscopic images of Heat Shield Rock, and then placed the Mössbauer spectrometer instrument on the rock for a 19-hour observation. Mössbauer integration times are longer now because the Mössbauer source has weakened as expected since landing.

On this sol in 210 (2005 January 14):

MER-A Sol 368: This was a remote sensing sol. *Spirit* made observations with its panoramic camera and its miniature thermal emission spectrometer and performed a successful test of the right eye of the panoramic camera to find the Sun. The rover team usually uses the left panoramic camera to locate the Sun.

Sun finding is sometimes called "get fine attitude" or "attitude update," and is something engineers do every couple of weeks to correct error in the rover's knowledge of attitude — mostly which way is north. This takes the same kind of images of the Sun that the atmospheric science team does, but the engineers use the data to determine attitude. Between the updates, the rover uses the onboard computer to keep track of attitude changes, but error builds up in this measurement over time. In general, most of the panoramic camera images of the Sun are acquired for atmospheric science. Many images are used to determine how much dust is in the atmosphere (atmospheric opacity or tau). Usually, engineers take these images three or four times during the day. With the current dust storms, the team is taking even more images of the Sun.

—NASA/JPL

221 Rishabha 02 **2025 September 24 07:48 UTC**
Ls 145.0° Curiosity Sol 4669
MSD 53938 Perseverance Sol 1634

On this sol in 210 (2005 January 14):

MER-B Sol 348: *Opportunity* placed the alpha particle X-ray spectrometer instrument on Heat Shield Rock during the afternoon for an overnight observation. That instrument provides best results when it is cold.

On this sol in 210 (2005 January 15):

MER-A Sol 369: Sols 369, 370, and 371 (210 Rishabha 02 through 04) were part of a three-sol plan. On Sol 369, *Spirit* looked for more science targets en route to Larry's Lookout.

—NASA/JPL

221 Rishabha 03 **2025 September 25 08:27 UTC**
Ls 145.5° Curiosity Sol 4670
MSD 53939 Perseverance Sol 1635

On this sol in 155 (1901 August 07):

Yuliya Ippolitovna Solntseva (Ю́лия Ипполи́товна Со́лнцева) was born; Russian/Soviet actress; starred in the motion picture *Aelita: Queen of Mars (Аэлита - королева Марса)* (1924).

On this sol in 210 (2005 January 15):

MER-B Sol 349: *Opportunity* brushed an area on Heat Shield Rock using the rock abrasion tool, took microscopic images of the brushed spot, then placed the alpha particle X-ray spectrometer on the spot for another overnight observation.

On this sol in 210 (2005 January 16):

MER-A Sol 370: *Spirit* looked for more science targets en route to Larry's Lookout.

—NASA/JPL

221 Rishabha 04 **2025 September 26 09:07 UTC**
Ls 146.0° Curiosity Sol 4671
MSD 53940 Perseverance Sol 1636

On this sol in 210 (2005 January 16):

MER-B Sol 350: *Opportunity* changed tools on the arm to the Mössbauer instrument for another long observation on the brushed area.

On this sol in 210 (2005 January 17):

MER-A Sol 371: *Spirit* completed a 6-meter (20-foot) drive to arrive at the rock target Peace. Due to the slope and small rocks in area, *Spirit* sat at an overall tilt of 19 degrees to the north-northeast, which was very good for maximizing solar energy.

—NASA/JPL

221 Rishabha 05 **2025 September 27 09:47 UTC**
Ls 146.5° Curiosity Sol 4672
MSD 53941 Perseverance Sol 1637

On this sol in 172 (1933 August 01):

"Castaways on Deimos" by J. Harvey Haggard appeared in *Wonder Stories* (date approximate).

Cover and interior art by Frank R. Paul.

On this sol in 210 (2005 January 17):

MER-B Sol 351: *Opportunity* changed tools on the arm back to the alpha particle X-ray spectrometer instrument again, putting it in place for another overnight observation of Heat Shield Rock.

On this sol in 210 (2005 January 18):

MER-A Sol 372: *Spirit* deployed the rover arm and acquired a set of images of Peace taken by the microscopic imager. A sequencing error prevented the alpha particle X-ray spectrometer from being placed on the rock, delaying the planned integration. The opacity of the sky, or tau, which is the amount of light that cannot penetrate through the atmosphere, rose sharply from 0.8 to 1.1.

—NASA/JPL

221 Rishabha 06 **2025 September 28 10:26 UTC**
Ls 147.0° Curiosity Sol 4673
MSD 53942 Perseverance Sol 1638

On this sol in 210 (2005 January 18):

MER-B Sol 352: *Opportunity* completed its work on Heat Shield Rock during Sols 347 through 352 (210 Rishabha 01 through 06), then got into position for more observations of the heat shield. This rock is now known to be an iron-rich meteorite, thanks to findings of the alpha particle X-ray spectrometer, Mössbauer spectrometer and the miniature thermal emission spectrometer. *Opportunity* remains in good health.

Opportunity took some final microscopic images of the rock then backed away for observations with the panoramic camera and miniature thermal emission spectrometer. *Opportunity* then drove to within about 1 meter (about 3.3 feet) of the largest piece of the heat shield in preparation for more observations of spent heat shield material with the microscopic imager.

The team continues to closely monitor orbital images for signs of dust storms. Tau, a measure of the sky opacity, has hovered in the 0.8 to 0.9 range for the past week; tau was roughly 0.5 before recent dust disturbances in the region.

On this sol in 210 (2005 January 19):

MER-A Sol 373: Solar energy continues to be a precious resource for *Spirit* because of the high tau on Sol 373. Although dust storms are more likely at this time of the marsrev, the start of the true dust storm season is still months away.

Spirit acquired more images with the microscopic imager and brushed Peace with the rock abrasion tool. The rover then placed the alpha particle X-ray spectrometer on Peace successfully for a nighttime integration. Tau continued upwards to 1.3, further reducing solar energy for *Spirit.*

Spirit is healthy, but reduced sunlight has been reaching the rover through the atmosphere due to a possible dust storm identified from orbital data. Despite limited energy during the period from Sol 367 through Sol 373 (210 Taurus 28 through Rishabha 06), *Spirit* made good progress by driving about 20 meters (66 feet) closer to top of Cumberland Ridge. *Spirit* is investigating a rock called Peace.

—NASA/JPL

221 Rishabha 07 **2025 September 29 11:06 UTC**
L_s 147.6° Curiosity Sol 4674
MSD 53943 Perseverance Sol 1639

On this sol in 210 (2005 January 19):

MER-B Sol 353: This was a restricted sol for *Opportunity,* as results of the Sol 352 (210 Rishabha 06) drive were not known by the planning team in time to calculate the final heat shield approach. *Opportunity* performed over two hours of observations using the panoramic camera and miniature thermal emission spectrometer.

On this sol in 210 (2005 January 20):

MER-A Sol 374: *Spirit* performed a successful 40-minute grind on Peace. The short grind time was due to energy limitations on the rover, which were a result of a dust storm that limited solar energy. The rock abrasion tool performed with no anomalies, and a 3.22-millimeter hole was made, removing some of the upper rock surface. Then, *Spirit* placed the alpha particle X-ray spectrometer in the hole drilled by the rock abrasion tool. The usual morning and evening remote-sensing science observations were conducted.

—NASA/JPL

221 Rishabha 08 **2025 September 30 11:45 UTC**

Ls 148.1° Curiosity Sol 4675
MSD 53944 Perseverance Sol 1640

On this sol in 152 (1897 November 06):

Frank Marvin Readick Jr. was born; American actor; starred in The Mercury Theatre on the Air's radio play "Invasion from Mars" (1938).

On this sol in 210 (2005 January 21):

MER-B Sol 354: *Opportunity* performed 10 minutes of pre-drive remote-sensing observations, then moved forward to get in final position for extending its instrument deployment device (robotic arm) to the heat shield. A drive of 0.7 meter (2.3 feet) was successful, placing the heat shield in reach of the arm. *Opportunity* performed more than an hour of post-drive imaging.

MER-A Sol 375: *Spirit* placed the Mössbauer spectrometer in the abraded hole. Again, the usual remote-sensing science observations were made.

—NASA/JPL

221 Rishabha 09 **2025 October 01 12:25 UTC**

Ls 148.6° Curiosity Sol 4676
MSD 53945 Perseverance Sol 1641

On this sol in 143 (1879 January 17):

Maxwell Karger was born; American film director; directed the motion picture *A Message from Mars* (1921).

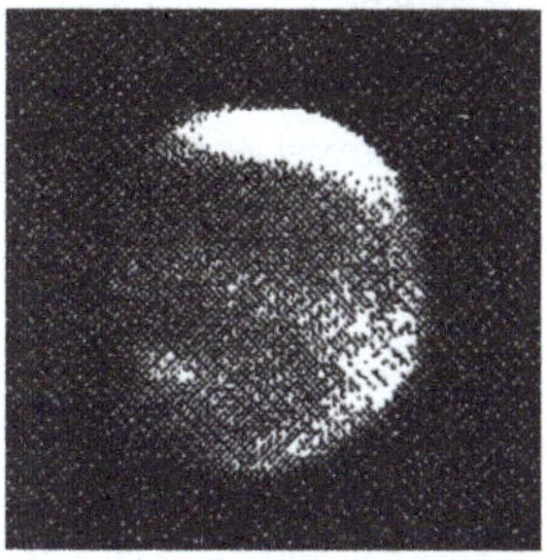

On this sol in 149 (1890 May 01):

Harvard University astronomers photographed Mars.

According to William H. Pickering, the Martian date was August 29.

On this sol in 187 (1961 October 20):

Augie Doggie and Doggie Daddy, Season 3 Episode 6, "Vacation Tripped" airdate. Augie and Doggie are planning their hunting vacation, but Augie wants to know how the hunting is on Mars.

On this sol in 210 (2005 January 22):

MER-B Sol 355: *Opportunity* performed over two hours of remote sensing during a restricted sol.

MER-A Sol 376: *Spirit* continued the Mössbauer spectrometer integration, then the rover changed tools to take pictures of the rock abrasion tool hole with the microscopic imager. *Spirit* performed more remote-sensing science.

—NASA/JPL

221 Rishabha 10 **2025 October 02 13:05 UTC**
Ls 149.1° Curiosity Sol 4677
MSD 53946 Perseverance Sol 1642

On this sol in 155 (1901 August 14):

George Worthing Yates was born; American screenwriter; co-wrote the screenplay adaptation from the book by Willy Ley and Chesley Bonestell for the motion picture *Conquest of Space* (1955).

On this sol in 210 (2005 January 23):

MER-B Sol 356: *Opportunity* used the microscopic imager to examine the heat shield. Using the arm to position the microscopic imager, *Opportunity* spent 90 minutes collecting high-resolution images of the heat shield.

MER-A Sol 377: *Spirit* completed a successful 1-hour-and-56-minute grind on Peace. This was the second attempt at grinding this rock target because the depth of first grind on Sol 374 (210 Rishabha 07) was deemed insufficient by the science team to really get inside the rock. The rock abrasion tool again performed the operation with no anomalies. It ground an additional 9.7 millimeters into the previously abraded surface. This additional depth did not necessarily add directly to the depth achieved on Sol 374 due to potential rover arm placement errors, but it was the deepest rock abrasion tool hole created to date.

—NASA/JPL

221 Rishabha 11 **2025 October 03 13:44 UTC**
Ls 149.6° Curiosity Sol 4678
MSD 53947 Perseverance Sol 1643

Mars arrival opportunity from Earth via a Type II trajectory.

On this sol in 155 (1901 August 15):

Frances Marie Deegan was born; American science fiction writer; author of "The Martian and the Milkmaid" (1944).

On this sol in 210 (2005 January 24):

MER-B Sol 357: *Opportunity* performed thermal inertia measurements throughout the sol. Using the miniature thermal emission spectrometer to image the same target at different times, *Opportunity* took measurements as late as 23:00 local time.

The *Opportunity* team celebrated *Opportunity's* first anniversary (one Earth year) on Mars. *Opportunity* continues to be in excellent health for its long drives out on the plains of Meridiani.

MER-A Sol 378: *Spirit* started the Mössbauer spectrometer reading at noon and integrated until 3:00 a.m. on Sol 379. The usual atmospheric science was conducted.

—NASA/JPL

221 Rishabha 12 **2025 October 04 14:24 UTC**

Ls 150.2° Curiosity Sol 4679
MSD 53948 Perseverance Sol 1644

On this sol in 210 (2005 January 25):

MER-B Sol 358: *Opportunity* retook some microscopic images of the heat shield with the dust cover open. *Opportunity* then stowed its arm and began its drive south, away from the heat shield. *Opportunity* is now headed for a small crater called Argo, which is approximately 300 meters (about 984 feet) away. *Opportunity* successfully covered 86.3 meters (283 feet) on this sol.

MER-A Sol 379: *Spirit* restarted the Mössbauer spectrometer to get four hours of observations during the day, and an alpha particle X-ray spectrometer reading was taken overnight. Atmospheric science, miniature thermal emission spectrometer readings of Boycott, and an alpha particle X-ray spectrometer reading were taken overnight.

—NASA/JPL

221 Rishabha 13 **2025 October 05 15:03 UTC**

Ls 150.7° Curiosity Sol 4680
MSD 53949 Perseverance Sol 1645

Mars Science Laboratory rover *Curiosity* is scheduled to begin its eighth marsrev of operations in Gale Crater.

On this sol in 210 (2005 January 26):

MER-B Sol 359: After spending 25 sols at the heat shield and nearby meteorite, *Opportunity* has completed its investigation of both and has started a long migration south. *Opportunity* is currently heading for a small crater called Argo. Dust storms in the vicinity of Meridiani Planum appear to be settling down, and solar power has stabilized.

MER-A Sol 380: *Spirit* had another productive week in the locale of a target called Peace. Because of strong interest in the unusual character of Peace, the team decided to throw the full science instrument payload at the rock.

The Science Operations Working Group (SOWG) decided that *Spirit* should remain at the location with Peace. *Spirit* collected pictures of the deep rock abrasion tool hole with the microscopic imager. The microscopic imager images of the bottom of that hole were received on the ground. A short alpha particle X-ray spectrometer integration on a new Peace target was collected just before the *Mars Odyssey* orbiter communications pass, and the data were sent down on that pass. A long alpha particle X-ray spectrometer integration (greater than eight hours) will come down later. Additional pictures of Peace from the microscopic imager were collected. The miniature thermal emission spectrometer monitored dust. The panoramic camera and hazard avoidance camera took images of the rock abrasion tool's magnets and grinding bit. The atmosphere has become slightly less dusty since the latest reporting.

—NASA/JPL

221 Rishabha 14 **2025 October 06 15:43 UTC**

Ls 151.2° Curiosity Sol 4681

MSD 53950 Perseverance Sol 1646

On this sol in 175 (1939 April 01):

"The Martian Avenger" by John Russell Fearn (as Polton Cross) appeared in *Amazing Stories* (date approximate). A baby of two worlds was this Martian, but even after thousands of years he remembered that revenge was his mission on Earth.

Interior artwork by Julian Krupa.

"Madness from Mars" by Clifford D. Simak appeared in *Thrilling Wonder Stories* (date approximate). A small, lonely Martian brought back to Earth unwittingly causes madness in humans and animals.

Interior art by H.W. Wesso.

"The Stronger" by Fredric Arnold Kummer Jr. appeared in *Dynamic Science Stories* (date approximate). Instead of a space-trip from Venus to Earth, Jim Weston found himself a shanghaied member of a slave-crew, Mars-bound on Saturnian Slane's rocket ship. And when they crashed on the red desert of Mars, this Earthman showed a Saturnian bully that on any planet its, guts, not size, that makes the man!

Interior art by Jack Binder.

On this sol in 210 (2005 January 27):

MER-B Sol 360: Since the *Opportunity* team was operating in restricted-sol mode, the team chose to plan Sols 360 and 361 together as a drive sol followed by a remote-sensing sol. On Sol 360, *Opportunity* traversed a record 154.65 meters (507.4 feet), using a combination of blind drives and auto-navigation software.

—NASA/JPL

221 Rishabha 15 **2025 October 07 16:22 UTC**

Ls 151.7° Curiosity Sol 4682

MSD 53951 Perseverance Sol 1647

On this sol in 210 (2005 January 28):

MER-B Sol 361: *Opportunity* carried out three hours of remote-sensing observations.

MER-A Sol 382: *Spirit* made a 4-meter (13-foot) approach to the next target, informally named Alligator. The drive succeeded as planned, putting Alligator perfectly in the workspace for rover-arm operations.

—NASA/JPL

221 Rishabha 16 **2025 October 08 17:02 UTC**

Ls 152.3° Curiosity Sol 4683
MSD 53952 Perseverance Sol 1648

On this sol in 210 (2005 January 29):

MER-B Sol 362: Sols 362 and 363 were planned together as another two-sol plan, again with the basic intent of driving as far as possible. After a directed drive of 90 meters (295 feet), *Opportunity* turned 180 degrees and continued in auto-navigation mode, resulting in an impressive 156.55-meter (513.6-foot) traverse. That is a new record for a single sol of driving on Mars. Alternating *Opportunity's* drive direction is part of the engineering strategy for maintaining the long-term health of our wheel drives.

MER-A Sol 383: *Spirit* performed a couple hours of remote-sensing observations.

—NASA/JPL

221 Rishabha 17 **2025 October 09 17:42 UTC**

Ls 152.8° Curiosity Sol 4684
MSD 53953 Perseverance Sol 1649

On this sol in 156 (1903 July 10):

John Wyndham Parkes Lucas Beynon Harris was born; British science fiction writer; author of *Exiles on Asperus* (1933), *Stowaway to Mars* (1936), "Time to Rest" (1949), "No Place Like Earth" (1951), "Dumb Martian" (1952), and "Mars A.D. 2094" (1958).

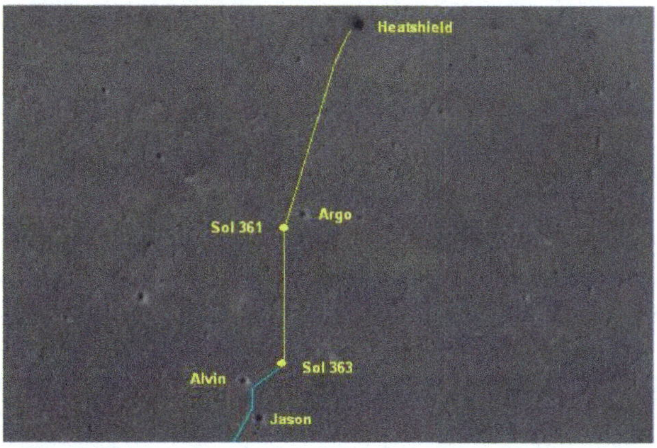

On this sol in 210 (2005 January 30):

MER-B Sol 363: *Opportunity* was commanded to continue driving for up to 120 meters (394 feet), as long as no drive errors had occurred on Sol 362 (210 Rishabha 16). However, due to a previously unidentified navigation software vulnerability, the Sol 363 drive errored out at its start.

MER-A Sol 384: *Spirit* performed a couple hours of remote-sensing observations.

—NASA/JPL

221 Rishabha 18 **2025 October 10 18:21 UTC**

Ls 153.3° Curiosity Sol 4685
MSD 53954 Perseverance Sol 1650

On this sol in 150 (1892 March 27):

Ferdinand Rudolph von Grofé was born; American composer, arranger, and pianist; composed the score for the motion picture *Rocketship X-M* (1950).

On this sol in 173 (1935 July 01):

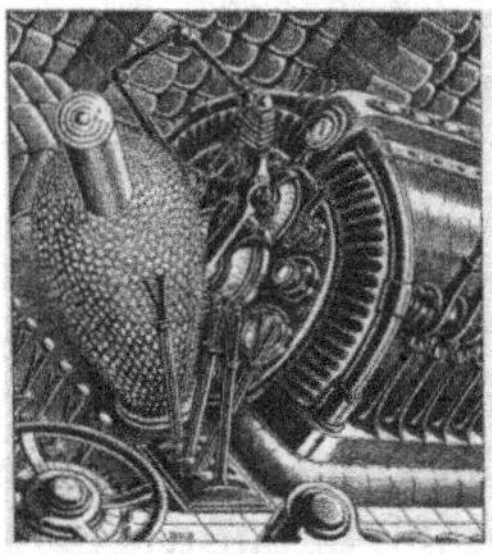

"The Son of Old Faithful" by Raymond Z. Gallun appeared in *Astounding Stories* (date approximate). In the second novella of the trilogy, the Martian's son races to Earth to help his father's friends forestall imminent invasion by his world's autocratic regime.

Interior art by Elliott Dold, Jr.

"The Far Way" by David R. Daniels appeared in *Astounding Stories* (date approximate). A mysterious recurrent situation; Bill Stuart is present for a time, then he suddenly disappears and is not seen for years. This pattern emerges in 1937, 1940, 1952, 1987, 2063, and on into the future. On each occasion Stuart has not changed noticeably in appearance. A mental superman of about A.D, 1,040,000 explains that Stuart's life-line, which should normally run through the fourth dimension, is warped by the sixth dimension, so that it skips long periods.

—Bleiler and Bleiler, 1998

Interior art by Marco Marchioni.

On this sol in 210 (2005 January 31):

MER-B Sol 364: The plan for *Opportunity's* Sol 364 was to continue the series of long traverses south. Unfortunately, there was a problem with a coolant line at a Deep Space Network transmitter, most of the pass was lost, and the plan could not be uplinked in the couple of minutes remaining. So, for what may have been the first time during *Opportunity's* mission, a sol's worth of nominal activities was lost, and the science run-out sequence from earlier commands was executed instead. Run-out sequences give *Opportunity* some useful tasks to do in case it does not get a new set of commands.

MER-A Sol 385: On Sols 385 and 386, *Spirit* completed complicated rover-arm operations. During 80 minutes on Sol 385, *Spirit* scoured some side-by-side patches of Alligator's surface with the wire bristles on the rover's rock abrasion tool, creating a brush mosaic. *Spirit* took readings with the Mössbauer spectrometer for 90 minutes, switched tools to the alpha particle X-ray spectrometer for 20 minutes of data-collection with that tool, then switched back to the Mössbauer spectrometer for more observations past midnight.

—NASA/JPL

221 Rishabha 19 **2025 October 11 19:01 UTC**

Ls 153.9° Curiosity Sol 4686

MSD 53955 Perseverance Sol 1651

On this sol in 210 (2005 February 01):

MER-B Sol 365: *Opportunity* used its panoramic camera and miniature thermal emission spectrometer to observe a target dubbed "Strange Rock," then moved a few meters (several feet) to get in position for trenching through a dune ripple crest.

MER-A Sol 386: *Spirit* continued Mössbauer spectrometer observations through 3:00 a.m. Mars local solar time. Later, *Spirit* took close-up pictures of the brush mosaic using the microscopic imager. *Spirit* then stowed the rover arm. The rover moved 1.2 meters (3.9 feet) back away from Alligator, and then performed an hour of post-drive imaging.

—NASA/JPL

221 Rishabha 20 **2025 October 12 19:40 UTC**

Ls 154.4° Curiosity Sol 4687

MSD 53956 Perseverance Sol 1652

On this sol in 210 (2005 February 02):

MER-B Sol 366: *Opportunity's* planners worked very closely with the science team to choreograph a rover trenching dance: *Opportunity* moved, scuffed (dragging its front wheels backwards multiple times), and finally trenched in the sand, all while placing the rover in a good orientation for later communications. The rover then went into deep sleep in preparation for an early morning photometric observation.

MER-A Sol 387: This was a restricted sol because information from the previous sol's operations could not be available in time for the team to plot further driving. So, *Spirit* spent 2.5 hours performing remote-sensing observations using the panoramic camera and miniature thermal emission spectrometer.

—NASA/JPL

221 Rishabha 21 **2025 October 13 20:20 UTC**

Ls 154.9° Curiosity Sol 4688

MSD 53957 Perseverance Sol 1653

On this sol in 210 (2005 February 03):

MER-B Sol 367: For *Opportunity's* Sols 367 and 368, a two-sol plan focused on investigation of a trench that the rover had dug with its wheels on Sol 366. *Opportunity* awoke on Sol 367 at about 7:30 a.m. local solar time after a night in the deep-sleep mode. It made some early-morning photometry measurements, then napped until the morning uplink window from 10:40 to 11:00 a.m. local solar time. After this, the rover acquired microscopic images of the trench wall, performed a short reading with the Mössbauer spectrometer, and then positioned the alpha particle X-ray spectrometer for data collection. After an afternoon communications relay session via *Mars Odyssey,* the rover slept until the sol 368 morning relay pass, at which time it started the alpha particle X-ray spectrometer integration.

MER-A Sol 388: *Spirit* has completed examination of a rock target called Alligator, using every tool on the instrument deployment device (robotic arm). With *Spirit's* batteries recharged and atmospheric dust stable again, the rover is in excellent health and ready to approach Cumberland Ridge, a crest on Husband Hill.

—NASA/JPL

221 Rishabha 22 **2025 October 14 21:00 UTC**
Ls 155.5° Curiosity Sol 4689
MSD 53958 Perseverance Sol 1654

On this sol in 156 (1903 July 15):

James Edwin McConnell was born; British book and magazine cover artist best known for Western and historical subjects; created the cover art for *The Gods of Mars* (1953), *The Warlord of Mars* (1953), *Thuvia, Maid of Mars* (1953), *A Fighting Man of Mars* (1954), *The Chessmen of Mars* (1954), and *The Master Mind of Mars* (1955).

On this sol in 210 (2005 February 04):

MER-B Sol 368: In the morning, *Opportunity* acquired more photometry observations, gathered more microscopic images, performed another short Mössbauer integration, and then positioned the alpha particle X-ray spectrometer for an overnight integration.

—NASA/JPL

221 Rishabha 23 **2025 October 15 21:39 UTC**
Ls 156.0° Curiosity Sol 4690
MSD 53959 Perseverance Sol 1655

On this sol in 140 (1873 June 10):

Spiritualist Elizabeth Denton conducted the second of her twelve "examinations" of Mars, in which she described a volcano and a mountain range green with tropical vegetation, large stone buildings, and polished marble slabs.

On this sol in 210 (2005 February 05):

MER-B Sol 369: *Opportunity* completed more trench investigations. It gathered additional microscopic images on new targets in the trench, completed another short data-collection session with the Mössbauer spectrometer, and placed the alpha particle X-ray spectrometer for another overnight integration.

MER-A Sol 390: *Spirit* drove closer to Larry's Lookout, about 13 meters (43 feet) backwards uphill. *Spirit* stopped when the rover reached the mobility time-of-day limit, which is a time of day that engineers program into the software to ensure the rover won't deplete all of its power at the end of a day's drive.

—NASA/JPL

221 Rishabha 24 **2025 October 16 22:19 UTC**

Ls156.5° Curiosity Sol 4691

MSD 53960 Perseverance Sol 1656

On this sol in 140 (1873 June 11):

Spiritualist Elizabeth Denton conducted the third of her twelve "examinations" of Mars:

> I see immense buildings. They seem to be dwellings, and are large enough to be the homes of thousands of people. Almost every thing I see here appears to be on a magnificent scale. One of the buildings is four or five stories high, with turrets and towers, and various designs both for ornament and convenience, which altogether relieve what might otherwise prove a disagreeable monotony. This building is square; that is, it encloses a square, containing, I think, a full square mile, -perhaps a mile and a half. The lower portion is divided into sections; and between each section and the one ad joining is a wide arched way, or opening.
>
> In the centre of the land thus enclosed stands a large circular building, much higher than this which encloses it ; and in it, in which she described immense dwellings capable of housing thousands of people, Surrounding a large building in which "there appears to be performed by machinery, and with almost incomprehensible speed and ease, a vast amount of that labor so common to every household, and which, in our isolated homes, constitutes, for so many mothers, wives, and daughters, the drudgery of life."
>
> At the base of this central building a wide, re volving platform extends entirely around it. From this platform a track, or road, runs in a straight line to every archway referred to as between the various sections of the large square building; and over these roads any person residing in a given section passes at will to any other section by simply seating himself in the conveyance which passes to and from that platform at regular intervals of only a few moments each ; then, stepping upon the platform until it re volves to the road corresponding to the section that he wishes to visit, he seats himself in a similar car, and is at his destination in a moment.
>
> The grounds enclosed, and corresponding to these various sections, are, in most instances, divided into plats of green similar to our grass, and beds of choice and fragrant flowers of almost every conceivable form and shade. The sparkling spray from fountains of pure water washes the flowers, cools the air, and beautifies every thing around. There are here blossoming and fruit-bearing trees, and shrubs and arbors and shaded walks, and very, very much that is beautiful beyond description. Many of the trees and plants resemble those of the tropics of our own planet.
>
> This style of building is no more common, I think, than a circular form, which is in every other respect quite similar. Both styles prevail extensively.
>
> I seem to be in one of the rooms now. It is very large. There is furniture here. I see long lounges, or seats, on which people recline. I obtain this by sensing the life of the people; but I do not see them, nor many of the objects in the room.

There is a peculiar air of scientific investigation here, -a great deal of thought, of study, an atmosphere of culture and intellectual strength. The per sons here are devoted to scientific research, and in a great many different directions. I feel all this and more; but I see no one.

I also feel the influence of great mechanical skill here; but it is subordinate to the scientific influence. The mechanical skill seems older and more perfect than the scientific ability. Both have, I think, reached greater perfection on this than on our own planet.

Such music I never heard. I have been trying to learn its source; but I cannot find it. It is clearer, sweeter, and at the same time more penetrating, than any I ever heard before.

On this sol in 210 (2005 February 06):

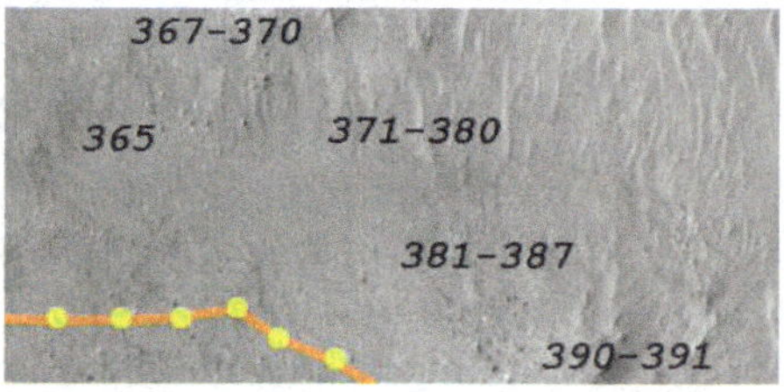

MER-B Sol 370: *Opportunity* acquired more microscopic images, more Mössbauer data and a variety of remote-sensing observations before using the deep-sleep mode overnight.

MER-A Sol 391: *Spirit* performed a set of remote sensing observations.

—NASA/JPL

221 Rishabha 25 | **2025 October 17 22:58 UTC**

Ls 157.0° | Curiosity Sol 4692

MSD 53961 | Perseverance Sol 1657

On this sol in 149 (1890 May 16):

Giovanni Schiaparelli drew a sketch of Mars. Mare Tyrrhenum, Hellas and Syrtis Major are identifiable.

Schiaparelli's sketch

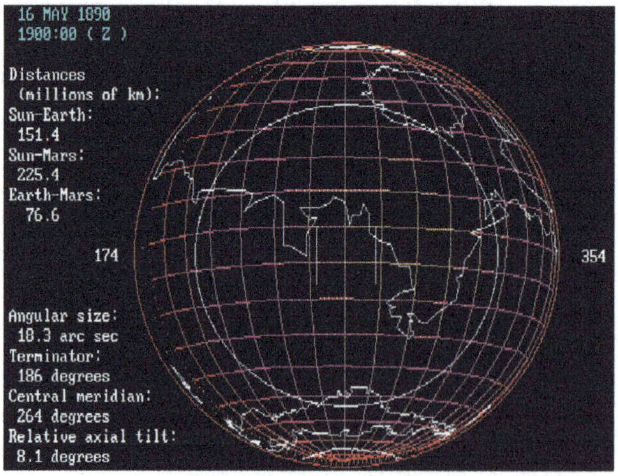

Mars's aspect, 19:00 Z

On this sol in 210 (2005 February 07):

MER-B Sol 371: After waking from deep sleep, *Opportunity* restarted the Mössbauer integration. *Opportunity* made remote-sensing observations during the middle of the day. Later, it gathered the last microscopic images on the trench, stowed its robotic arm and used its left front wheel to scuff the soil. *Opportunity* then bumped backwards to put the scuffed area into the arm's work volume.

MER-A Sol 392: This sol was planned as a 23-meter (75-foot) drive toward Larry's Lookout, plus some post-drive imaging in the drive direction with the navigation camera and panoramic camera. The usual remote-sensing science was planned for various times throughout the sol. However, *Spirit* halted after completing the first segment of the drive, a distance of only about 12 meters (39 feet). The halt may have been due to rocks nearby considered too dangerous by the rover's autonomous navigation system. This left *Spirit* with more driving to do on the next drive opportunity. *Spirit* also has been taking energy from the batteries recently (due to increased dust in the atmosphere, which has covered the solar panels with a thin layer of dust, blocking some of the light that provides energy via the solar panels).

—NASA/JPL

221 Rishabha 26 **2025 October 18 23:38 UTC**

Ls 157.6° Curiosity Sol 4693
MSD 53962 Perseverance Sol 1658

Earth departure opportunity to Mars via an EEM trajectory.

Mars arrival opportunity from Earth via a Type III trajectory.

On this sol in 140 (1873 June 13):

Spiritualist Elizabeth Denton conducted the fourth of her twelve "examinations" of Mars, in which she described a light-skinned utopia and a dark-skinned dystopia. Of the first race:

> The people seem to be in advance of all I have seen on the earth in many respects. They are physically and mentally our superiors. They seem to have based their physical culture upon scientific principles, until they have reached the very perfection of physical development. They have studied to live in harmony with all the laws of their physical life, until they appear to have no temptation to violate those laws. The result is a wonderful purity, harmony, and completeness of the whole physical being.
>
> In their social life there is more leisure, more order, more individuality, than among us. They seem to live the idea expressed by the motto, "Each for all, and all for each," and to have done this until they know no narrow, selfish ends in life, and have no desire for aggrandizement, honor, or preferment, that is not wholly subservient to the public good. Indeed, they seem to have controlled selfishness and all the baser passions, until they are the mere servants of the higher and nobler powers of the soul. I question whether they have what we recognize by the word governments, except it may be such regulations as they deem necessary for the more effective distribution of means for the accomplishment of desired ends. And in this there is not, that I can now discover, any antagonism, and never discord. Their controversies,

if they have them, seem only for the discovery of truth, never for triumph of person or party. What we recognize as courts of justice are to them unknown or unemployed institutions. I sense no such difference in social conditions as we have here: I do not think they exist. I cannot find among them any persons who seem low or degraded; nor do I see such suffering as we meet almost everywhere on this planet.

Physically they do not appear to be as heavy as we. There is a buoyancy in this respect that I do not understand. The complexion is of a purer white than is that of the Anglo-Saxon race; the eyes are generally of a bright sky-blue; and the hair varies in shade from a bright straw to that of a deep gold color. Mentally they seem able to accomplish any thing they undertake. They have studied our planet, and have become very familiar with it. I believe they have traced us out as a people. I do not know how they have done this; but they seem to know a good deal about us. They seem familiar with other planets also. I believe we can, in time, learn to hold communication with them. Perhaps they are psychometers!

They appear to have a great deal of time for culture; and, besides, they are such thinkers, and such workers! They make actual conditions conform to ideal conditions. All share in the delights, and all lend ready and willing aid to secure them.

Of the second race:

Now I am in a place a long way from there, where all is different. The scenery is changed; every thing is changed: but the great difference is in the people. These are more like the people of this planet; that is, more like the majority of our people here. They seem like an entirely different race from those I saw before. I cannot see that these are separated from the others by any great natural division of the surface of the planet; but their condition, their habits, and their lives, are wholly different.

I get mental conditions best. I find antagonisms and selfishness that I did not find with that other people. All that I see here are dark colored, very dark; but I should not call them black. I cannot get individual features. I think there is a difference in physical structure; but I cannot get it clearly. They seem much like us in many respects. There seems to me to be a world-wide difference between us and the other people I saw. It will take us a long time to get where they are, even physically. I do not like to go among these people from the others. I do not think they know about the others. Yes; they know about them: but they do not seem to have the least idea that they can ever be like them; nor do they care to be like them. They differ greatly from us in this respect. They seem to feel with regard to those superior people as the Indians do with regard to us. Taken as a people, they are lower in the scale of intelligence than are many of the peoples of this planet.

On this sol in 176 (1941 March 01):

"The City of Mummies" by Edgar Rice Burroughs appeared in *Amazing Stories* (date approximate). John Carter takes a solitary pleasure flight to the dead city of Horz, only to find it isn't as empty as he thought. The last descendants of the Orovars, the earliest race on Barsoom, capture him after he comes to the aid of one of their warriors, Pan Dan Chee. The Orovars guard knowledge of their existence, and so condemn Carter and the man he rescued to death. They send him and Pan Dan Chee into the pits to await execution. In the pits, they discover a living dead hypnotist, Lum Tar O, who has placed men and women from the earliest days of Mars into hypnosis and kept them in crypts. Carter also discovers hiding in one of the sarcophagi his granddaughter, Llana. She explains that an arrogant jeddak, Hin Abtol of the northern people of Panar, kidnapped her, but she brought his flier down in Horz and took shelter in the tombs.

—Ryan Harvey

Cover and interior art by J. Allen St. John.

"Star of Blue" by Milton Kaletsky appeared in *Science Fiction* (date approximate). Marshall Hughes, leader of Earth's fighting men, finds himself forced to surrender to the heartless ruler of Mars—until he looks into the heavens where the fate of his world rests of the identification of a color.

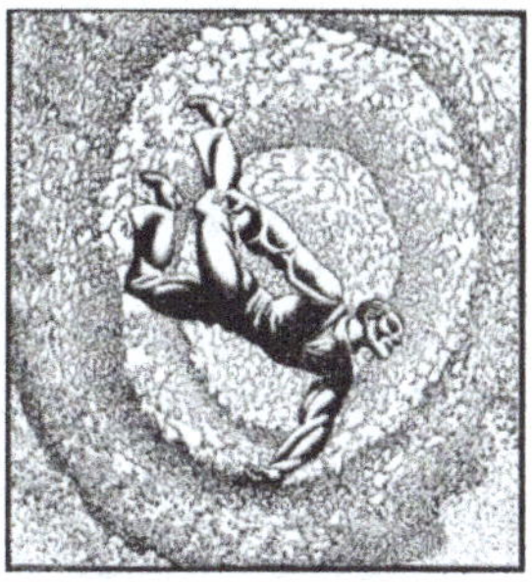

"The Secret Sense" by Isaac Asimov appeared in *Cosmic Stories* (date approximate). Lincoln Fields, a rich Earthling from New York City living on Mars, is discussing the merits of the highly developed sense of sight and hearing that humans possess, which the Martians do not. Garth Jan, his Martian friend, counters with the merits of the senses that the Martians possess, and unintentionally lets slip that there is also one secret sense. Fields asks him to give it to him but Garth Jan refuses at that time. Six months later, using Martian social law in an unethical fashion, Fields forces him to let him feel this sense and Jan reluctantly agrees.

—Wikipedia

Interior art by Hannes Bok.

"The Martians Are Coming" by Robert W. Lowndes also appeared in *Cosmic Stories* (date approximate). When the inebriated experimenters invited the Martians to come to Earth, they didn't really mean it, but when the Martians took them at their word and sent a thousand armed ships…

Interior art by David A. Kyle.

“Death Walks in Washington” by James Norman Schmidt (as James Norman) appeared in *Fantastic Adventures* (date approximate). Martians are a funny looking shape (akin to a penguin) with a tulip shaped nose and do not communicate in sound with each other, but through odors. It is this talent that lends itself to Oscar being able to solve many of the Earth’s mysteries when he is magically transferred from Mars to Earth by the magician, Hodar. Through their adventures, Oscar, assisted by Hodar, solve mysteries using his nose and marvelous deductive reasoning power.

—Barnesandnoble.com

Interior art by Julian S. Krupa

“Blockade Runner” by Malcolm Jameson appeared in *Astounding Science-Fiction* (date approximate). A piece of Campbellian science fiction in which Earth and Mars are at war and the Martians have set up a blockade so that a critical fuel supply cannot be transported to Earth’s fleet.

Interior art by Charles Schneeman.

On this sol in 210 (2005 February 08):

MER-B Sol 372: *Opportunity* completed microscopic imaging of the scuffed area, collected Mössbauer data, and switched to the alpha particle X-ray spectrometer. *Opportunity* did not use the deep sleep mode overnight so that it could perform an overnight reading with the alpha particle X-ray spectrometer.

On this sol in 210 (2005 February 09):

MER-A Sol 393: *Spirit’s* solar panels were collecting a fine layer of dust, which had reduced energy levels, but *Spirit* kept on keeping on.

The rover team planned an easy remote-sensing day in order to try to put back some energy into the batteries.

—NASA/JPL

221 Rishabha 27 **2025 October 20 00:18 UTC**
Ls 158.1° Curiosity Sol 4694
MSD 53963 Perseverance Sol 1659

On this sol in 149 (1890 May 19):

An astronomer identified by William H. Pickering as “Guillaume” (possibly Guillaume Bigourdan) drew a sketch of Mars. Tyrrhenum Mare, Syrtis Minor, and Syrtis Major are identifiable.

According to the Pickering calendar the Martian date was August 47.

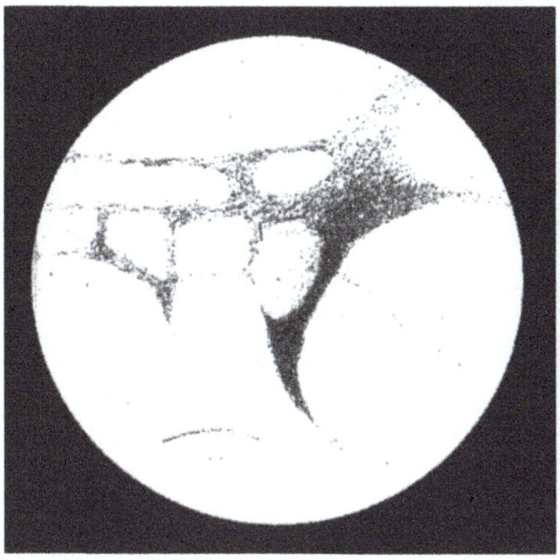

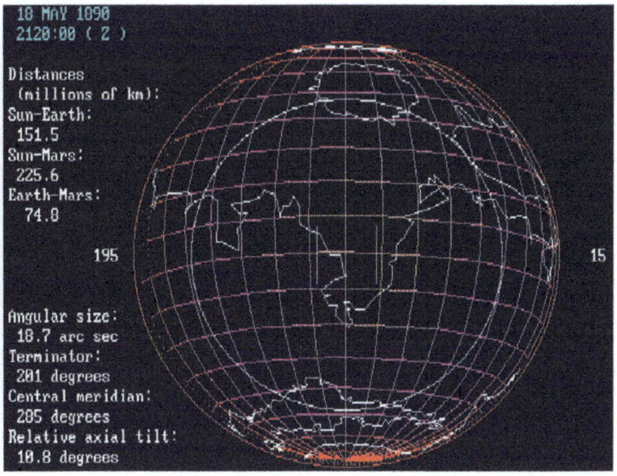

Guillaume's drawing | Mars's aspect, 21:20 Z

On this sol in 210 (2005 February 09):

MER-B Sol 373: *Opportunity* completed the alpha particle X-ray spectrometer work and then changed tools back to the Mössbauer for more integration during the day. In the afternoon *Opportunity* acquired some final microscopic images of the scuff, used the hazard avoidance camera to inspect wear on the grinding teeth of the rock abrasion tool, and then stowed the arm. *Opportunity* then bumped back about 1.5 meters (about 5 feet) to position itself for observing the trench with the instruments on the mast. It turned to a heading of 250 degrees to be in good position for four hours of high-gain antenna tracking for receiving an upload of improved flight software.

On this sol in 210 (2005 February 10):

MER-A Sol 394: *Spirit* drove to a safe spot for good communications with Earth during upload of new software.

—NASA/JPL

INDEX

Made in United States
Orlando, FL
05 November 2026

90209476R00155